EUROPEAN ROAD LIGHTING TECHNOLOGIES

PREPARED BY THE STUDY TOUR TEAM

Dale Wilken FHWA	Paul J. Lutkevich Parsons Brinckerhoff	John Arens FHWA
Balu Ananthanarayanan Wisconsin DOT	C. Paul Watson Alabama DOT	Jim Havard LITES
Patrick Hasson FHWA	Karl Burkett Texas DOT	Jeff Unick Pennsylvania DOT

and

American Trade Initiatives, Inc.
&
Avalon Integrated Services, Inc.

for the

Federal Highway Administration
U.S. Department of Transportation

and

The American Association of State Highway and
Transportation Officials

and

The National Cooperative Highway Research Program
(Panel 20-36)
of the Transportation Research Board

September 2001

FHWA INTERNATIONAL TECHNOLOGY EXCHANGE PROGRAMS

The FHWA's international programs focus on meeting the growing demands of its partners at the Federal, State, and local levels for access to information on state-of-the-art technology and the best practices used worldwide. While the FHWA is considered a world leader in highway transportation, the domestic highway community is very interested in the advanced technologies being developed by other countries, as well as innovative organizational and financing techniques used by the FHWA's international counterparts.

INTERNATIONAL TECHNOLOGY SCANNING PROGRAM

The International Technology Scanning Program accesses and evaluates foreign technologies and innovations that could significantly benefit U.S. highway transportation systems. Access to foreign innovations is strengthened by U.S. participation in the technical committees of international highway organizations and through bilateral technical exchange agreements with selected nations. The program has undertaken cooperatives with the American Association of State Highway Transportation Officials and its Select Committee on International Activities, and the Transportation Research Board's National Highway Research Cooperative Program (Panel 20-36), the private sector, and academia.

Priority topic areas are jointly determined by the FHWA and its partners. Teams of specialists in the specific areas of expertise being investigated are formed and sent to countries where significant advances and innovations have been made in technology, management practices, organizational structure, program delivery, and financing. Teams usually include Federal and State highway officials, private sector and industry association representatives, as well as members of the academic community.

The FHWA has organized more than 40 of these reviews and disseminated results nationwide. Topics have encompassed pavements, bridge construction and maintenance, contracting, intermodal transport, organizational management, winter road maintenance, safety, intelligent transportation systems, planning, and policy. Findings are recommended for follow-up with further research and pilot or demonstration projects to verify adaptability to the United States. Information about the scan findings and results of pilot programs are then disseminated nationally to State and local highway transportation officials and the private sector for implementation.

This program has resulted in significant improvements and savings in road program technologies and practices throughout the United States, particularly in the areas of structures, pavements, safety, and winter road maintenance. Joint research and technology-sharing projects have also been launched with international counterparts, further conserving resources and advancing the state of the art.

For a complete list of International Technology Scanning topics, and to order free copies of the reports, please see list on the facing page.

Website: www.international.fhwa.dot.gov
Email: international@fhwa.dot.gov

FHWA INTERNATIONAL TECHNOLOGY EXCHANGE REPORTS

Infrastructure

Geotechnical Engineering Practices in Canada and Europe ⌂
Geotechnology-Soil Nailing ⌂
International Contract Administration Techniques for Quality Enhancement-CATQEST ⌂

Pavements

European Asphalt Technology ⌂⌂
European Concrete Technology ⌂⌂
South African Pavement Technology
Highway/Commercial Vehicle Interaction
Recycled Materials in European Highway Environments ⌂

Bridges

European Bridge Structures
Asian Bridge Structures
Bridge Maintenance Coatings
European Practices for Bridge Scour and Stream Instability Countermeasures
Advanced Composites in Bridges in Europe and Japan ⌂
Steel Bridge Fabrication Technologies in Europe and Japan ⌂
Performance of Concrete Segmental and Cable-Stayed Bridges in Europe ⌂

Planning and Environment

European Intermodal Programs: Planning, Policy and Technology ⌂
National Travel Surveys ⌂
Recycled Materials in European Highway Environments ⌂
Geometric Design Practices for European Roads ⌂

Safety

Pedestrian and Bicycle Safety in England, Germany and the Netherlands ⌂
Speed Management and Enforcement Technology: Europe & Australia ⌂
Safety Management Practices in Japan, Australia, and New Zealand ⌂
Road Safety Audits—Final Report ⌂
Road Safety Audits—Case Studies ⌂
Innovative Traffic Control Technology & Practice in Europe ⌂
Commercial Vehicle Safety Technology & Practice in Europe ⌂
Methods and Procedures to Reduce Motorist Delays in European Work Zones ⌂

Operations

Advanced Transportation Technology ⌂
European Traffic Monitoring
Traffic Management and Traveler Information Systems
European Winter Service Technology
Snowbreak Forest Book – Highway Snowstorm Countermeasure Manual (*Translated from Japanese*)
European Road Lighting Technologies ⌂

Policy & Information

Emerging Models for Delivering Transportation Programs and Services
Acquiring Highway Transportation Information from Abroad—Handbook ⌂
Acquiring Highway Transportation Information from Abroad—Final Report ⌂
International Guide to Highway Transportation Information ⌂

⌂ **Also available on the internet**

⌂⌂ **Only on the internet at www.international.fhwa.dot.gov**

CONTENTS

OVERVIEW viii

INTRODUCTION 1
   Trip Planning 1
      Objective 2
      Team Members 2
      Meetings 2
      Amplifying Questions 3
      Trip Itinerary 3
   Report Organization 4
   American and European Contrasts 4
      Culture 4
      Language 5
      Engineering 5

PRACTICAL MATTERS OF ROADWAY LIGHTING SYSTEMS 6
   Design 6
   Verification 7
   Equipment Quality Level and Lighting System Maintenance 7
   Power Conservation 8
   Master Lighting Plan 9
   Energy-Absorbing Poles 9
   Litigation 10
   Light Pollution 10
   Warrants 11
      Belgium 11
      Switzerland 11
      Finland 11
      France 11
      The Netherlands 11
   Panel Recommendations 13

VISIBILITY DESIGN 14
   Design 14
   Research 14
   Crosswalks 16
   Panel Recommendations 17

LUMINANCE DESIGN AND PAVEMENT REFLECTION FACTORS 19
   Luminance Design Technique 19
   Pavement Reflection Factors and R-Tables 19
   New Types of Pavement 21
   Pavement Reflection Factors: Other Conditions 22
   Panel Recommendation 23

TUNNELS 24

COUNTER-BEAM AND PRO-BEAM LIGHTING 28

CONTENTS

HIGH-MAST, DECORATIVE, AND SIGN LIGHTING 30

High-Mast Lighting 30
Decorative Lighting 30
Sign Lighting 30
Panel Recommendation 33

ROUNDABOUTS 34

Roundabout Categories 34
Luminaire Locations 34
Roundabout Light Levels 35
Panel Recommendation 35

SAFETY IMPLICATIONS 36

Other Observations 38
Panel Recommendation 39

FUTURE DEVELOPMENTS 40

New European Standards 40
Traffic Control Centers 40
Dynamic Roadway Lighting 41
Guidance Systems 42
Pavement Reflection Qualities 43
Tunnels 44
Research Needs 45
Overall Research Impressions 45
Panel Recommendations 45

SUMMARY OF RESEARCH RECOMMENDATIONS 46

ACKNOWLEDGMENTS 47

APPENDIX A: PANEL MEMBERS 48

APPENDIX B: AMPLIFYING QUESTIONS 53

APPENDIX C: KEY CONTACTS IN HOST COUNTRIES 60

APPENDIX D: KEY PAPERS 64

APPENDIX E: OUTREACH ACTIVITIES IN 2000 65

ENDNOTES 66

TABLES

1. Schedule of Team Meetings 3
2. Scan Team Itinerary 3
3. R-Table Values, by Pavement Class 20
4. Fatalities in Road Accidents 36
5. Fatalities per 1 billion vehicle kilometers traveled 37

CONTENTS

FIGURES

1. The roadway lighting scan team 2
2. Maintenance of luminaires, Switzerland 7
3. Tunnel cleaning in Paris 8
4. & 5. Master lighting plans, Paris 9
6. Results of crash test of energy-absorbing pole 9
7. & 8. Results of R-Tech's study on light pollution 10
9. Uniform vs. nonuniform lighting 15
10. Typical three-dimensional target 15
11. Model roadway installation 16
12. View of Lecocq's computer modeling software 16
13a. Demonstration roadway with 3-D spheres and square, flat targets 17
13b. Photographic image (zoom) of targets 17
13c. Synthesized image of targets 17
14. Synthesized configuration of road surface 18
15. Synthesized configuration of road surface 18
16. Lighting scheme for crosswalks, Switzerland 18
17. Wevelgem Tunnel, Belgium 19
18. Highway near Helsinki Airport, Finland 19
19. Highways near Helsinki, Finland 19
20. Milchbuck Tunnel, Switzerland 21
21. New porous asphalt 21
22. Porous asphalt, after 12 months 22
23. Wet roadway in Finland 22
24. Underground roundabout, Switzerland 24
25. Underground roundabout entrance and exit feeds to underground parking 24
26. Tunnel in Lyon, France 25
27. Tunnel in Helsinki, Finland 25
28. Tunnel at Schipol Airport, the Netherlands 26
29. Milchbuck Tunnel, Switzerland 26
30. Wevelgem Tunnel, Belgium 26
31. Milchbuck Tunnel, Switzerland 27
32. Black Window method, the Netherlands 27
33. Examples of Black Windows 27
34a. Symmetrical (bisymmetrical) light distribution 28
34b. Counter-beam light distribution 28
34c. Pro-beam light distribution 28
35. High-mast lighting, Finland 30
36. High-mast lighting, Belgium 30
37. Parking lot lighting, Helsinki, Finland 31
38. Parking lot lighting, Helsinki Airport, Finland 31
39. To aid recognition, vertical and semispherical illuminance is used in pedestrian areas 31
40a., b. & c. Example of typical decorative lighting in Zurich, Switzerland 32
41. & 42. Downtown Helsinki, Finland 32
43. & 44. Decorative lighting in Finland 33
45. & 46. Zurich, Switzerland, at night 33

47. Micro-prismatic sheeting materials for signs in Finland 33
48. Roundabout, Paris, France 34
49. Roundabout at Philips Outdoor Lighting Application Center,
    La Valbonne, France 34
50. Swiss recommendation for luminaire placement 35
51. The Swiss "Vision Zero" program 38
52. Road accidents compared with numbers of vehicles 39
53. & 54. Views of a TCC in Switzerland (left) and Finland (right) 40
55. Low level of roadway lighting, the Netherlands 41
56. Normal level of roadway lighting 41
57. High level of roadway lighting 41
58. Dutch guidance systems under investigation 42
59a. & b. In-road, fiber-optic delineators 43
60. Examples of colored pavement 43
61. & 62. Application of colored pavement 44
63. Typical motorist's view of a tunnel 44
64. Virtual reflectometer, France 45
65. Effects of tunnel lighting color 45

OVERVIEW

The volume of vehicle traffic is increasing worldwide, and roadway lighting can be an effective tool to help provide efficient and safe traffic movement. The U.S. transportation community is interested in identifying cutting-edge research and technologies in highway and roadway lighting systems. Specific interests include tunnel illumination, sign lighting, and visibility metrics that are used in the design of roadway lighting systems.

The American Association of State Highway and Transportation Officials (AASHTO) is in the process of updating its *Informational Guide for Roadway Lighting* and recognizes the need to gather information from transportation ministries and lighting professionals outside the United States. The information gathered will provide a basis to update the *Guide* and will provide a better tool for State and local authorities that design, install, operate, and maintain public lighting systems.

The study was co-sponsored by the U.S. Federal Highway Administration (FHWA), an agency of the U.S. Department of Transportation, and by AASHTO. The purpose of the study was to gather information related to current roadway lighting practices and innovative solutions used by other countries.

The team members brought a variety of professional perspectives to the study. Representation included the States of Alabama, Pennsylvania, Texas, and Wisconsin; the FHWA; and the Illuminating Engineering Society of North America (IESNA).

The lighting study was conducted during the first 16 days of April 2000, with meetings held in Finland, Switzerland, France, Belgium, and the Netherlands. The delegation met with professionals in the field of roadway lighting to observe and evaluate the European experience in a number of areas of specific interest, including small target visibility (STV) and luminance design techniques.

Information was collected on the following 10 primary areas of interest, which form the main sections of this report:

- Practical Matters
- Visibility Design
- Luminance Design
- Pavement Reflection Factors
- Tunnels
- Counter-Beam vs. Pro-Beam Lighting
- High-Mast Lights and Signs
- Roundabouts
- Safety Implications
- Future Developments

Based on its observations, the panel developed specific recommendations for the roadway lighting and safety communities in the United States. The recommendations appear below, in descending order of priority.

VISIBILITY DESIGN TECHNIQUE

The team members found that none of the countries visited use visibility techniques in design. Visibility research with three-dimensional targets is, however, being conducted in France and Belgium.

European research suggests that the visibility concept may provide a more complete approach to lighting design, although more experience is needed. The panel recommends experimentation and research on active roadways.

DYNAMIC ROAD LIGHTING

In the Netherlands, highway engineers have installed a dynamic roadway lighting system that can be operated at three levels, depending on the amount of traffic and weather conditions. The high level is 2 cd/m^2, the normal level is 1 cd/m^2, and the low level is 0.2 cd/m^2. The crash rate for the 0.2-cd/m^2 system, when operated at low traffic volumes, was acceptable. From these results it was determined that new systems will be installed to operate at 1 cd/m^2 and 0.2 cd/m^2. A similar road is currently being installed in Finland.

The French are studying retroreflectivity and active luminous devices. Similarly, the acceptability of different types of guidance systems is being researched in the Netherlands.

As an approach to more dynamic management of roadway lighting, the panel recommends investigating the application of dimmable lighting systems, turning off lighting systems, and alternative guidance systems.

PAVEMENT REFLECTION FACTORS

All of the countries use the luminance design method for roadways. Several countries noted that there are problems with the standard "R-tables." The initial luminance values measured in the field vary from the values predicted by the design calculations that used the standard R-tables.

It was stressed to the panel that, when doing field measurements, the roadway must be dry and the temperature must be above the dew point. It was also noted that better correlation between calculated and measured values is obtained when measurements are made in the summer.

The French are researching the photometric properties of road surfaces. The evolution of road surface technology and the use of bright and colored road surfaces necessitated the research. Examples of new road surfaces are "quiet" and "water-draining" pavements and very thin asphaltic concretes and surface dressings. The French also are examining the possibility of using a virtual reflectometer for field measurements.

Pavement reflectance is an important element of lighting design. The panel recommends that more research, including field measurements, be conducted in order to overcome the acknowledged inadequacy of the R-tables for pavements.

OVERVIEW

MASTER LIGHTING PLAN

A number of European cities have master lighting plans. The plans are based on providing safety, beautification, and security for goods and people. Urban lighting is viewed as a key component of city management.

The panel encourages the development of master lighting-design plans to improve the coordination of roadway and urban lighting in such matters as lighting levels and styles and themes for safety, security, and beautification.

ROUNDABOUT LIGHTING

Each of the countries visited has specific recommendations for roundabout lighting. All cited the importance of having roundabout light levels higher than the levels on approach streets.

The panel recommends that the European experience in roundabout lighting be synthesized and consolidated for U.S. application.

CROSSWALK AND PEDESTRIAN-AREA LIGHTING

The Swiss have modified lighting techniques to provide vertical illuminance, which allows pedestrians in crosswalks to be seen in positive contrast. The result has been a lowering of fatalities by two-thirds. Other countries also cited the importance of vertical illumination in pedestrian areas to enhance easy identification.

The panel recommends the consideration of vertical illuminance as a design approach to improve safety in crosswalks and other pedestrian areas.

ENERGY-ABSORBING POLES

Energy-absorbing poles flatten upon impact, but do not break away. They are used mainly in Finland and may be useful in the United States, in areas where breakaway poles are not desirable.

The panel recommends investigating the use of energy-absorbing poles as an option for selected applications.

EXPERIMENTATION

Throughout the trip, the team encountered many instances in which the Europeans gained knowledge and experience by conducting practical experiments on active roadways. This method permits more rapid implementation of new ideas.

The panel encourages more innovative experimentation on active roadways and test tracks.

CRASHES AND LIGHTING

The police in Zurich, Switzerland, presented an extensive accident report. The panel found it interesting that the police analyze the cause of automobile accidents in the Zurich area and make recommendations for lighting applications.

The panel recommends the development of reporting systems that consider the lighting conditions at crash scenes.

EUROPEAN LIGHTING STANDARDS

There is a potential to gather a great deal of information from European lighting documents. The panel recommends further evaluation of the European standards and guidance documents to determine applicability in the United States.

EQUIPMENT QUALITY LEVEL AND MAINTENANCE

The European lighting equipment generally appeared to be of a high quality, and very few roadway lighting outages were observed. The lighting systems were generally relamped on a group basis, typically on a 3- to 5-year cycle. Maintenance of tunnel lighting systems is generally conducted on a shorter cycle that coincides with the cycle for washing. Necessary relamping is conducted at that time. It was stated that the tunnels on the loop, in Paris, are cleaned every month.

The panel recommends that, when possible, higher quality lighting materials be considered to benefit maintenance and durability for the life of the lighting systems. In addition, maintenance personnel should be thoroughly trained to ensure the integrity of lighting systems.

SIGNS

Several countries are beginning to eliminate sign lighting by using micro-prismatic sheeting material. France also is moving away from fixed sign lighting by using engineering-grade retroreflective material.

The panel recommends the use of micro-prismatic materials for unlighted overhead and left-shoulder mounted signs.

INTRODUCTION

Vehicular travel is increasing throughout the world, particularly in large urban areas and at all hours of the day and night. At night, the visual capabilities of humans are impaired and visibility is reduced. Road crashes at night are disproportionately high in numbers and severity, compared with daytime crashes. In the United States, the nighttime fatality rate, weighted for kilometers traveled, is three times the daytime figure.[1,2] One of the major factors contributing to the problem is darkness, because of its influence on a driver's behavior and ability. Thus, roadway lighting can be an effective tool to help ensure efficient and safe traffic movement. The U.S. transportation community is interested in identifying cutting-edge research and technologies in highway and roadway lighting systems, including tunnel illumination, sign lighting, and all the methods that are used in the design of roadway lighting systems.

The American Association of State Highway and Transportation Officials (AASHTO) is in the process of updating its publication *Informational Guide for Roadway Lighting* and recognizes the need to gather information from transportation ministries and lighting professionals around the world. The information gathered could provide a basis on which to update the *Guide*, thereby providing a better tool for State and local authorities that design, install, operate, and maintain public lighting systems.

Recognizing the benefits that could result from an examination of international practices, a team of roadway lighting and safety experts was assembled. The team's mission was to observe and document practices that might have value to the U.S. transportation community. In April 2000, the panel traveled to five European countries (Finland, Switzerland, France, Belgium, and the Netherlands) to observe innovative lighting practices and identify those practices that could be implemented in the United States. This report describes the findings and observations of the group and includes recommendations of practices that have potential for implementation in the United States.

TRIP PLANNING

In 1990, the Federal Highway Administration (FHWA), in coordination with AASHTO and the Transportation Research Board (TRB), began an international transportation technology research program. The program involves assembling teams of experts in specific areas of transportation technology who travel overseas to identify technologies and practices that might have immediate or near-term implementation value in the United States. The cost of sending a group overseas and documenting the findings is significantly less than the cost of researching the technologies and preparing the appropriate documentation in the United States. In addition, individual team members benefit from firsthand observation of the technology applications in a real-world setting.

A scan trip begins when FHWA and AASHTO identify the need to observe international practices in a particular field. A panel of experts in that field is created, and the panel meets to plan the key aspects of the trip and develop a series of "amplifying questions" that are submitted to the host countries in advance of the trip. During the trip, panel members meet as a group with representatives of various

INTRODUCTION

organizations in each host country. Upon its return, the panel prepares a report describing its observations and recommendations.

Objective

The objective of this study was to review and document European experience with roadway lighting systems and advanced technologies, such as small target visibility (STV) and counter-beam technologies, in tunnels and roadways and for special geometries such as roundabouts. Findings may be incorporated in the new AASHTO *Informational Guide for Roadway Lighting*, which is due for revision in the near future. The scan team also set out to observe innovative technologies that may be implemented in the United States in the near or long term.

The study panel also was interested in aspects of planning, installation, operation, maintenance, and financing, as they relate to innovative lighting systems. In gaining an understanding of innovative lighting systems and technologies, the panel hoped to identify both the similarities and differences between European and U.S. systems that might affect implementation. The panel also wanted to identify problems associated with implementing innovative technologies and systems and the role(s) that nongovernment, private entities had in implementing and operating lighting systems. Finally, the panel wished to observe, firsthand, the systems and technologies in operation and obtain information to assess their effectiveness.

Team Members

The team members represented several different perspectives including that of the FHWA, four State departments of transportation (Alabama, Pennsylvania, Texas, and Wisconsin), and the Illuminating Engineering Society of North America (IESNA).

Appendix A lists the panel members, their affiliations, and short biographies. Figure 1 shows the panel members during their visit to France.

Meetings

The panel met four times throughout the trip development and the actual tour, as shown in table 1. The first meeting provided an opportunity to define the areas of greatest interest and prepare a series of amplifying questions that the host countries could use to develop programs for presentation to the team.

Figure 1. The roadway lighting scan team: from left, Paul Watson, Jim Havard, Paul Lutkevich, Karl Burkett, Balu Ananthanarayanan, John Arens, Marie-Dominique Gorrigan (Alt), Dale Wilken, Pat Hasson, and Jeff Urick.

TABLE 1. SCHEDULE OF TEAM MEETINGS.

Location	Date and Time Frame	Purpose
Washington, D.C.	1/13/00	Determine emphasis areas and develop amplifying questions
Helsinki, Finland	4/2/00 (Beginning of tour)	Plan trip actions and emphasis areas
Lyon, France	4/9/00 (Mid-tour)	Review findings
Utrecht, The Netherlands	4/16/00 (End of tour)	Identify key findings and develop preliminary panel recommendations

Amplifying Questions

To provide the European hosts with a clearer understanding of the issues and technologies of interest, the team prepared a series of amplifying questions that focused on 10 major topics, as listed below:

- Future Developments
- Practical Matters
- Visibility Design
- Luminance Design
- High-Mast, Decorative, and Sign Lighting
- Tunnels
- Pavement Reflection Factors
- Counter-Beam vs. Pro-Beam Lighting
- Roundabouts
- Safety Implications

The amplifying questions are listed in appendix B.

Trip Itinerary

The tour took place during the first two weeks of April 2000. Table 2 lists the countries and cities visited by the study panel.

TABLE 2. SCAN TEAM ITINERARY.

Dates	Countries	Cities
April 3 – 4	Finland	Helsinki
April 5 – 7	Switzerland	Zurich
April 10	France	Lyon
April 11	France	Paris
April 12	Belgium	Liege
April 13 – 14	The Netherlands	Utrecht

Note: Only the dates on which the panel members met with hosting officials are listed. The table does not include travel days and weekend panel meetings.

Appendix C lists the officials with whom the panel met during the trip. The hosts presented information on a wide variety of lighting topics, and the panel observed many other interesting practices during the tour. Many of the hosting agencies provided documents to the scanning team. The documents referred to in this report are listed in appendix D.

REPORT ORGANIZATION

During the tour, the panel identified many noteworthy practices, several of which may have current or future value to transportation agencies in the United States. Each section of this report begins with a brief description of the topic, then documents the panel's observations, and concludes with a recommendation. The final section contains a summary of the panel's research recommendations. Appendix E lists opportunities for the team members to share the information at conferences and through technical articles and demonstrations.

AMERICAN AND EUROPEAN CONTRASTS

Throughout the tour, team members were continually educated on some of the significant differences between the United States and the European countries visited. The differences were evident in many areas, including culture, language (both common and technical), and engineering practices. While the engineering differences were the focus of the trip, the other differences affected the gathering of information and also will have an impact on the ability of U.S. practitioners to implement promising technologies or practices.

Culture

Although the focus of the trip was on innovative lighting systems, panel members had the pleasure of experiencing the people and facilities in each country. As they traveled on planes, trains, subways, buses, and taxis, stayed in different hotels, and interacted with the people in each country, the panel members were able to observe many significant cultural characteristics in the five countries. Many cultural characteristics represent nothing more than a different way of living and give each area its unique identity. Some cultural characteristics, however, have a direct impact on the lighting systems in each country. Many of the cities visited have very dense, active populations that engage in extensive walking or bicycling. Comprehensive trolley and subway systems are used for both work and recreation. Additionally, many automobiles compete for the limited parking. Also, large numbers of people were out and about in the center of town at night. The team members surmised that this nighttime activity prompted the local governments to light buildings, parks, and monuments for the users' comfort and security, as well as for display.

Preservation of urban centers is important to Europeans. As a result, Europeans have a very strong sense of history and the preservation of that history. The antiquity and historical importance of European cities is a magnet for tourists and of great economic importance.

Generally, Europeans also appear to have great respect for authority, which leads to high compliance with traffic-control regulations and devices. In many cases, the panel identified practices that were innovative or unique, but that would have limited application in the United States because of basic differences in lighting systems and cultures.

Language

The panel members were continually impressed by the ability of their hosts to communicate in English. The majority of individuals the panel met with were fluent in English. Even so, the panel had to learn numerous terms, both common and technical. A few of the most common are listed below, with the European term listed first and the equivalent American term in parentheses.

- Motorway (freeway)
- Carriageway (travelway or paved roadway)
- Dual carriageway (divided highway)
- Hard shoulder (paved shoulder)
- Columns (poles)
- Junction (intersection)
- Lorry (truck)
- Petrol (gasoline)
- Control gear (ballast)

Engineering

It was evident to the team members that their European counterparts have had many of years of experience with designing solutions and managing lighting problems in cities and rural areas on all classes of roadways. The panel found many solutions practical, effective, and, more often than not, new and creative. European engineers are utilizing new technologies faster than many of their U.S. counterparts, and European transportation agencies appear to be more progressive in testing and implementing new technologies and applications of lighting systems. The difference may be due, in large part, to the aggressive and progressive research programs in the individual countries. Many of the solutions observed on roadways were certainly more advanced than those that are used on roadways in the United States. Examples include the use of variable lighting levels, depending on time of day, weather, and traffic movement; traffic guidance systems, in place of fixed, overhead lighting systems; energy-absorbing poles, in areas where frangible poles could not be used; master lighting plans to guide long-term development; and vertical illumination in crosswalks.

One of the most significant engineering contrasts is the Europeans' willingness to gain knowledge and experience by conducting practical experiments on active roadways. This method permits rapid implementation of innovative ideas. In defense of the lack of experimentation in the United States, Europeans do not experience the amount of litigation that regularly occurs in the States. Therefore, in Europe, it is easier to do actual research on public roads.

PRACTICAL MATTERS OF ROADWAY LIGHTING SYSTEMS

In the area of practical matters, the panel was interested in examining details involved with design, verification, operation, and maintenance of European roadway lighting systems.

DESIGN

Generally, lighting is installed in Europe at a higher light level than is used in the United States, and the roadway lighting is more uniform in appearance. The higher lighting levels and more uniform appearance are the results of many studies over the years that examined visual performance and visual comfort. Additionally, the panel heard the Europeans equating higher light levels with driver comfort, which, they believe, produces a higher level of safety. (It should be noted, however, that numerous studies have been conducted, the results of which have not been conclusive. French experts pointed out that while higher light levels contribute to driver comfort, they may also create a false sense of safety, masking drivers' levels of fatigue or intoxication.)

In Europe, the luminance design method is widely used for standard road sections. The illuminance design method is used for more complex situations such as intersections, pedestrian crossings, roundabouts, residential areas, rest areas, and bicycle-path lighting. Currently, each country visited has its own design standards that are based on the documents produced by the Commission Internationale de l'Eclairage (CIE). Among countries the lighting levels are approximately the same for equivalent classes or types of roadways. In addition, each country has developed its own guide for lighting designers to consult. The guides address matters of mounting height, spacing, overhang, lamp wattage, lamp type, and type of luminaire.

Of all the countries visited, only Switzerland is not a member of the European Union (EU). However, all are working through the Comité Europeén de Normalisation (CEN), which is the European Committee for Standardization, to produce harmonized lighting standards that will apply to all EU members.

It was common to find the design process outsourced to contractors. In Finland, the government had negative experiences with performance specifications, because contractors reduced installations to the minimum limit acceptable, which reduced long-term suitability. To solve the problem, the typical specification is written, based on experience, to read "manufacturer, catalog number, or similar."

In Belgium, most (80 percent) of the motorways (freeways) are continuously lighted, for safety reasons. The traffic intensities on Belgian motorways are very high and the distances between interchanges are short (each 3 to 4 km) because of the high degree of urbanization. In less densely populated areas, only the interchanges of the motorways are lighted.

As noted earlier, in the Netherlands, motorway light levels have been reduced to approximately the same range of values as those used in the United States. The uniformity of lighting is, however, still typically European. The Dutch seem to be satisfied with the lower light levels.

Finnish representatives mentioned that they have ceased using low-pressure sodium (LPS) as a light source, primarily because of the cost of the lamp.

VERIFICATION

Rather than testing individual luminaires for tunnel applications, field measurements of lighting levels are conducted on the majority of tunnels in Switzerland. Roadways, however, are measured only if there appears to be a problem. (In Finland, calculations are verified, but no field verification is currently conducted.) Because lighting contractors cannot be held accountable for road surfaces, verification is usually done in lux. The French designs are based on luminance and verified by measuring illuminance (lux).

EQUIPMENT QUALITY LEVEL AND LIGHTING SYSTEM MAINTENANCE

The panel reviewed some of the European lighting equipment and, in general, concluded that it was of a higher quality level than that generally available in the United States. In addition, the team noted that the relamping maintenance of the lighting systems was very good (see figure 2). Typically, the road lighting systems are relamped on a group basis, on a 3- to 5-year cycle.

The French are experimenting with remote control and monitoring of public lighting systems. Monitoring data will include time of operation, proper operation, automatic troubleshooting, and problem notification. Control includes on/off control and possible future dimming. Fixtures employ electronic high-pressure sodium (HPS) ballasts.

Figure 2. Maintenance of luminaires, Switzerland.

As the team traveled around the five countries at night, it had many opportunities to observe each country's lighting. Generally, the lighting was better maintained than comparable lighting in the United States. The team rarely observed unlit luminaires and was impressed with the overall uniformity and quality of the lighting.

The panel noted that Europe shares a problem with the United States, i.e., matching existing photometrics, or overall lighting performance, on an existing system. In discussions with representatives in various countries, it became evident that the Europeans have not achieved an effective means of maintaining the photometric performance of the lighting systems. Once a system is designed and built, no systematic lighting measurements are made in the long term and no controls are placed on replacement luminaires. This causes a rapidly deteriorating performance of the system. The scope of effort required to correct the problem and the cost involved are enormous, which has prompted the Europeans not to require maintenance of initial performance levels.

In Belgium, the regions or cities are responsible for the installation and maintenance of lighting installations; contractors are hired on a low-bid basis to perform the installation and maintenance. No controls or contract requirements are made to maintain photometric performance of the lighting systems. Contractors are responsible for obtaining fixtures and other replacement items. Contractors do not typically have a lighting engineer or a lighting-design expert on staff and have little incentive to maintain the photometric performance of existing systems.

In-depth discussions with the Belgians on maintenance issues confirmed that their concerns are similar to some U.S. concerns. These concerns are that maintenance personnel are not capable of determining the photometrics of the existing system and that they are not able to determine acceptable alternatives. Contractors typically replace luminaires with whatever is in stock. It is difficult to write and enforce specifications for photometrics for replacement fixtures that would provide equivalent luminance values.

The French allow the installing contractor to select fixtures to meet a luminance calculation specification. French representatives stated that contractors do not usually understand the luminance design method and pay little attention to it. Field measurements of the built system are made with illuminance values. Maintenance personnel do not attempt to match replacement fixture photometric performance to that of initial fixtures.

The problem is more critical on lighting systems designed for lower light levels, where poles are spaced farther apart. On such systems, nearby fixtures will have a higher percentage contribution at each point, and distant fixtures will have a lower percentage contribution, making fixture photometric performance increasingly important for maintenance of the system performance.

Figure 3. Tunnel cleaning in Paris.

Tunnel lighting system maintenance is done on a shorter cycle, approximately two to four times per year, which coincides with the common washing and spot relamping cycle. The panel noted that all the tunnels on the loop in Paris are cleaned every month, as shown in figure 3.

POWER CONSERVATION

The Swiss Energy Administration has a standard, not a law, on the lighting density limit (watts/m^2) that sometimes affects the design light levels. The Administration also limits the amount of annual energy consumption (KW h/yr) for lighting. To meet the requirements, some lighting is reduced in the late night, typically from 11 PM to 5 AM.

Although France has no limits on power consumption, it is not unusual to dim the lighting, to save energy, between the hours of 10 PM and 6 AM. A recent survey by the Center for Studies on Urban Planning, Transport, Utilities, and Public Constructions (CERTU) shows that one-third of French towns decrease lighting at night, and 8 percent of the networks are dimmed at night.

In Finland, an analysis of lighting-system costs over 20 years shows that electric energy is two-thirds of the total cost. To save energy, some Finnish roadways have high/low-style controls, and light levels are lowered. The motoring public has not complained.

MASTER LIGHTING PLAN

Throughout the scan, the panel noted that a number of cities had developed formal master lighting plans. The plans accounted for economic and cultural changes, the public image of the city, and technological developments. The benefits of such a plan are that it organizes the different functions of lighting, plans the different parts of the city, and schedules the expenditures. For example, the City of Paris has developed a master urban lighting plan that is based on the safety of the roads, beautification, and security for goods and people (figures 4 and 5). Lighting is employed to change the image of the environment, re-link different parts of the city, and indicate the nature of the site. Essentially, planners consider urban lighting to be one component of managing the city.

Figures 4 & 5. Master lighting plans, Paris.

ENERGY-ABSORBING POLES

CEN has developed a new standard for breakaway and energy-absorbing poles, number EN12767, "Passive safety of support structures for road equipment." New types of poles meeting the standard and suitable for wind speeds of up to 23 m/s (approx. 50 mi/h) have been installed. Figure 6 shows an example of how the energy-absorbing pole works. The panel

Figure 6. Results of crash test of energy-absorbing pole.

thought that this item would have applicability in areas where it is not currently advisable to use breakaway poles.

Additionally, the panel was shown several bored-center-hole methods used in Finland to make wooden poles break away when impacted by a vehicle.

Crash testing of roadside devices is based on CEN standard procedures. The CEN procedures are directly modeled after the U.S. crash-test criteria specified in National Cooperative Highway Research Program (NCHRP) Report 350, with modifications; for example, pickup trucks are not included in European tests.

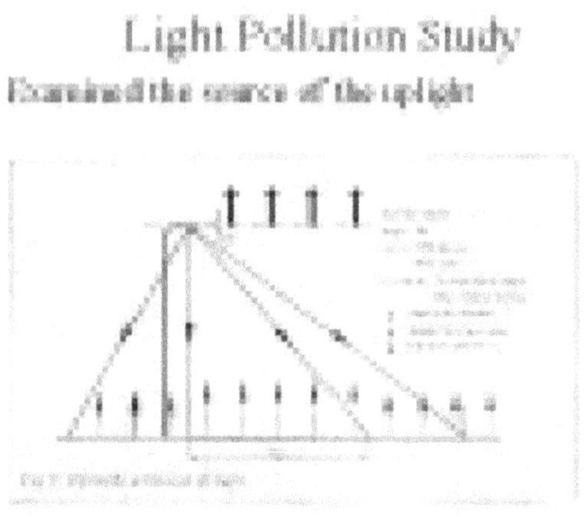

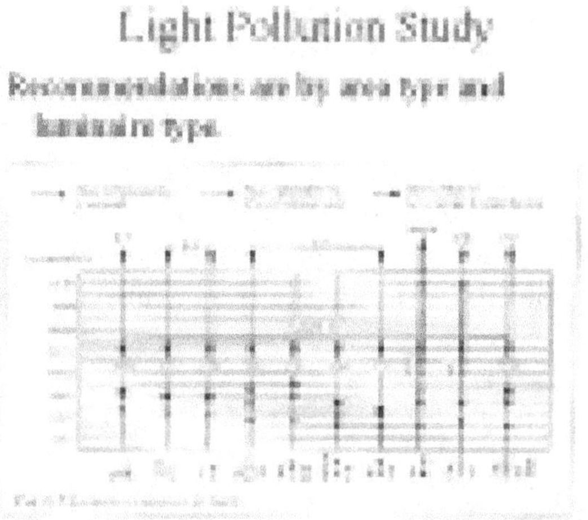

Figures 7 & 8. Results of R-Tech's study on light pollution.

LITIGATION

Litigation about lighting did not appear to be an issue in Europe, as it is in the United States, which may be explained, in part, by differences in the legal systems. The European approach also is different. For example, the Dutch have developed a national lighting policy that includes dimming. It was pointed out to the team: "If you follow the policy, no one can sue when you turn the lights out."

LIGHT POLLUTION

The team was not made aware of any formal sky glow restrictions in Europe. There is certainly, however, an awareness of the problem as well as a technical report, *Guidelines for Minimizing Skyglow*, CIE 126, 1997. Results of several studies by R-Tech in Belgium on the amount of uplight generated by various types of luminaires were presented to the panel in both France and Belgium and are shown in figures 7 and 8. In later feedback the panel was told that, in Belgium, the upward light ratio of the luminaires is limited. We also were informed that the light pollution figures are being updated and that this information will be available in the fall of 2001. To reduce the total amount of upward flux, the following guidelines have to be followed:

1. Reduce upward light output ratio (ULOR) as much as possible.

2. Maximize the utilization factor (K) in such a way that it should approach the downward light output ratio (DLOR) as much as possible.

3. In the case of a roadway lighting designed in luminance, maximize the luminance efficiency expressed in cd/m²/lux.
4. In the case of an outdoor installation designed in illuminance, apply strictly the requested illuminance level.

WARRANTS

Each country visited had different warrants, as noted below.

Belgium

A large proportion of paved roadways in Belgium are lighted. The continuous lighting of the motorways between interchanges has a luminance level of 1 cd/m² in the Flemish region and 1.5 cd/m in the Wallonia region; the interchanges are lit to a luminance level of 1.5 cd/m².

Switzerland

The official practice on Swiss federal highways (motorways) is to light only junctions (intersections). Highways in urban areas are lighted in the neighborhood of lighted urban roads or in areas with higher risks. The normal roads (mixed traffic) are mainly lighted in urban areas.

Finland

In Finland, warrants are very detailed, and safety is used as a justification for the installation of lighting. One of the special reasons cited for lighting two-lane roadways was the existence of an adjacent, lighted pedestrian way or lighted bike path. Approximately 20 percent of the roads in Finland are lit.

France

In France, responsibilities for lighting and its maintenance vary according to the "owner" of the roadway. The national government is responsible for national roads and motorways, counties are responsible for county roads, and cities for city roads. Individual towns are responsible for the lighting of national and county roads within an urban area. National roads must be lighted as follows:

> 50,000 Average Daily Traffic (ADT)	General lighting
25,000 to 50,000 ADT with interchanges < 5 km apart	General lighting
25,000 to 50,000 ADT with interchanges > 5 km apart	Interchange only
< 25,000 ADT	Interchange only

The Netherlands

As in France, individual towns in the Netherlands establish their own lighting policies. At the national level, specific warrants were written in 1990, referred to as the "1990 Warrants." The warrants to install lighting are as follows:

Always Light:

- Four or more lanes
- Dual carriageway with 1,500 vehicles/hour/lane peak
- Single carriageway with 2,000 vehicles/hour/lane peak
- Single carriageway with 1,800 vehicles/hour/lane peak, if slow-moving vehicles are present

Since 1997, an additional assessment is required, as follows:

1. Does the road meet the 1990 Warrants?
2. Does the road go through or near a scenic area?
    a. If no, then install lighting
    b. If yes, then move to No. 3
3. Conduct a supplementary study to consider lighting alternatives.
    a. Can glare screens be used?
    b. Is guidance lighting possible?
    c. Can the lighting be switched or dimmed?
    d. Is lighting acceptable, in view of cost/benefit?
4. Reach a conclusion. If the decision is to install lighting, install extra measures such as the ability to dim or turn off during late night.

As of the team's visit, a new lighting policy for highways was being written. It contains the following elements:

1. Traffic Safety: only enough light for traffic safety.
2. Energy Efficiency: goal of 10 percent improvement by 2010; not trading lives for energy consumption.
3. Environmental Consequences: (see above additional 1997 assessment).
4. Effects on Road Capacity: based on research at the Technical University of Delft, addition of standard lighting shows a 4 percent improvement. (Currently, the Dutch studies do not include the effects of dynamic lighting on road capacity.)
5. Administration and Maintenance of the Roadway.

Environmentalists in the Netherlands have raised consciousness about potential impacts of lighting on animal behavior. Humans, too, are included in the debates – some people want to live where it is dark at night. The efforts of environmentalists are changing the lighting policy.

PANEL RECOMMENDATIONS

- Develop master lighting design plans to improve the coordination of roadway and urban lighting in such matters as lighting levels, styles, and themes for safety, security, and beautification.

- Investigate the use of energy-absorbing poles as an option for selected applications. Careful investigation should be made into the adequacy of these devices, considering the wide range of vehicle mass and speed on U.S. highways.

- When possible, consider quality lighting materials to benefit maintenance and durability for the life of the lighting systems.

- Thoroughly train maintenance personnel to maintain the integrity of the lighting systems.

VISIBILITY DESIGN

The IESNA recently approved a revision to its publication RP-8, *American National Standard Practice for Roadway Lighting*. The revision includes three methods for designing continuous lighting systems for roadways: illuminance, luminance, and STV.

DESIGN

One of the primary reasons for conducting the European study was to meet with leading experts in the field of roadway lighting to find out about their experiences with using a visibility design metric.

While the panel found that a lot of research is being done in the area of visibility, none of the research has yet been implemented into everyday practice. In more than one country, team members heard the words, "We have no practical experience," when it came to applying the visibility design techniques.

Because of a negative experience, the Swiss have changed their approach to lighting crosswalks. They used to shine lights directly across the crosswalk, but discovered that, when at the curb, the pedestrian was less visible because the background varied, from buildings in some places to darkness in others. The Swiss now light crosswalks from the side, so the pedestrian is highlighted in positive contrast. Later input from the French confirmed the Swiss approach, but included the caveat that the main risk is that pedestrians often believe they are seen by drivers whatever the light distribution and weather conditions, even if they are not in the zebra marking.

The panel was shown roadways in Finland that appeared to have relatively nonuniform lighting. It was thought these lighting systems might provide a higher visibility level. Subsequent calculations by a team member have, indeed, shown that this road exceeded any luminance and STV requirements in the new ANSI/IESNA RP-8-00. Naturally, the ultimate measure of the quality of this type of design will be the change in the number of crashes. Crash data were not available at the time of the visit.

RESEARCH

The panel was pleased to notice the amount of visibility research being done in both France and Belgium. Team members saw spatial frequency analysis by Fast Fourier Transforms being used by several people (Eric Dumont, Philippe Boogaerts, etc.) to describe information content of a scene or border (edge) contrast. Additionally, Mr. Boogaerts indicated that the Fast Fourier Transform is also used in the processing of the images of charge coupled device (CCD) cameras. Both countries have selected the visibility model and equations of Dr. Werner Adrian and are using three-dimensional targets. Representatives in both countries told the panel that the visibility concept provided a more complete approach to lighting design and supplemented the information provided by the luminance approach that is commonly used throughout Europe. The approach, which utilizes three-dimensional targets, results in very uniform lighting (see figures 9 and 10).

The use of three-dimensional targets by the French and Belgians provides contrasts within the target, thereby making the target more visible.

Figure 10 shows a typical three-dimensional target used by the Belgians to develop simulation software. The Belgians[8] found good correlation between panel ratings and STV calculations for 20 percent targets, but stated that "the visual task of a driver cannot be considered as detection, within a useful time, of unexpected small static targets." They further stated, "The use of STV assumes full use of (factors) affecting visibility and knowledge of the limitations of the concept." They were adamant about the need to include headlights into the calculation of visibility.

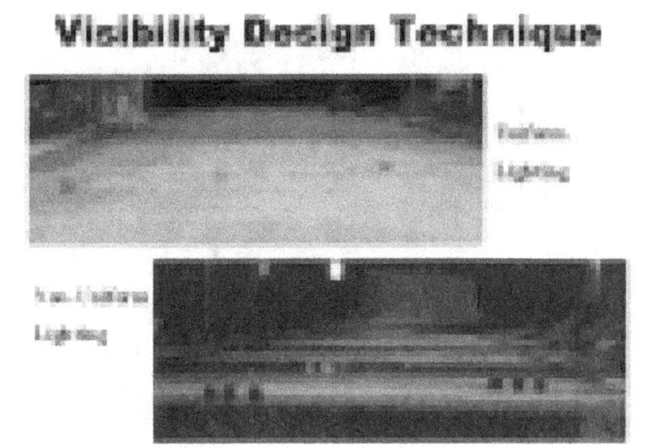

Based on extensive research done by Jacques Lecocq, the French have proposed that a simple minimum target visibility level (VL) metric is all that is necessary. Mr. Lecocq's research is based on translating a model roadway into a computer program that allowed many observer trials and the rapid collection of data (figure 11). The panel noted that Mr. Lecocq's model relied on approximations of key factors. These factors include the use of Lambertian distribution calculation of light reflected from the pavement and the shadow effect of multifaceted targets. Mr. Lecocq noted that, as targets get larger, the visibility always becomes greater. Large targets develop contrasts within themselves, as opposed to small targets, which are always viewed against their background, i.e., the roadway surface.

Figure 9. Uniform vs. nonuniform lighting.

Figure 10. Typical three-dimensional target.

The nine possible target positions are shown in figure 12, a view of Mr. Lecocq's software. The software permitted experimentation that determined the minimum visibility level needed for adequate lighting.

Based on an R2 roadway, a 0.35-s observation time, and a 20 percent reflectance target, the results of the study indicate that a minimum visibility level of 7 is needed for good visibility.

Using a ray-tracing computer program called "Radiance," the Belgians have developed synthesized computer targets that replicate real-world, illuminated, three-dimensional targets on a demonstration roadway, as shown in figures 13a, b, and c.

Figure 11. Model roadway installation.

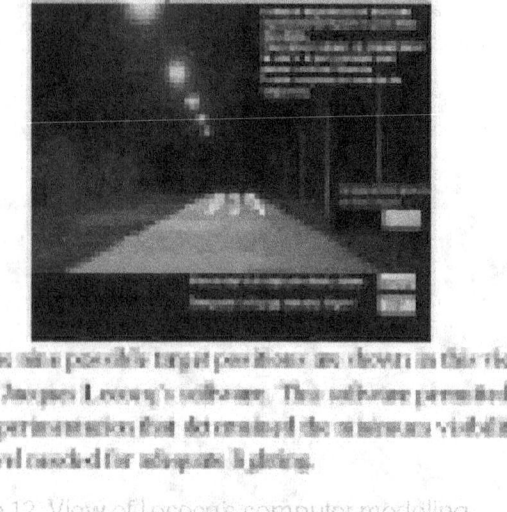

Figure 12. View of Lecocq's computer modeling software.

Studies utilizing the synthesized images have shown excellent correlation between the calculated levels of visibility and the subject assessments of the observers for both flat 20-cm x 20-cm and spherical targets (appendix D).

In addition, the Belgians believe that their work shows that good uniformity on a poorly lighted (<1 cd/m^2) road is insufficient. While the VL does improve as the roadway becomes more nonuniform, they believe that Belgian drivers would not accept the appearance of the roadway (figures 14 and 15). Later input from Mr. Lecocq further clarified that this increase in VL only applies to the average of several targets in the sense of mean values. Further, if one considers one target at a time whose reflection factor is variable, a flat one can be made visible or invisible simply by choosing an appropriate reflection factor. The flat target can even play the role of a type of specious amplifier for the average of individual VLs. This is generally not the case for a spherical target. On a roadway with poor longitudinal uniformity, typically all targets are either very visible or invisible depending on location. In this case, however, a corresponding mean value is not related to the ability to see any obstacle at any place on the road at a given time by the driver.

The Belgians[4] have found that headlights impact VL and should be included in calculations. Also, they believe that the VL approach is not usable in cluttered environments, i.e., environments with off-roadway sources, such as towns and residential areas. Therefore, the VL approach should be limited to the lighting of main roadways in rural areas.

Finally, the Belgians noted that, with the addition of the visibility design approach, lighting engineers are no longer limited to "producing luminance," but can also "produce visibility."

CROSSWALKS

The Swiss recently enacted a law recognizing that the pedestrian has the right of way in a crosswalk. The initial result of the new law was an increase in vehicle-pedestrian crashes. If circumstances in Switzerland are similar to those in the United States, the vast majority of pedestrian fatalities occur after dark. The Swiss studied the crosswalks and have based new crosswalk- and roundabout-lighting recommendations on the visibility principle of highlighting objects in positive contrast. As shown in figure 16, poles are positioned so that pedestrians are seen in positive contrast, when light levels are below 2 cd/m². No special pole positioning is required for light levels at or above 2 cd/m². Installation of the new lighting resulted in a two-thirds reduction in pedestrian-vehicle crashes, but an increase in minor vehicle-vehicle crashes, typically "rear-enders," resulting from quick stops.

PANEL RECOMMENDATIONS

- European research suggests that the visibility concept may provide a more complete approach to lighting design, though more experience is needed. The panel recommends experimentation and research on active roadways.

- The panel recommends the consideration of vertical illuminance as a design approach to improve pedestrian safety in crosswalks and other pedestrian areas. It also recommends research into the relative benefits between positive- and negative-contrast lighting techniques and development of appropriate levels.

Figure 13a. Demonstration roadway with three-dimensional spheres and square, flat targets.

Figure 13b. Photographic image (zoom) of targets.

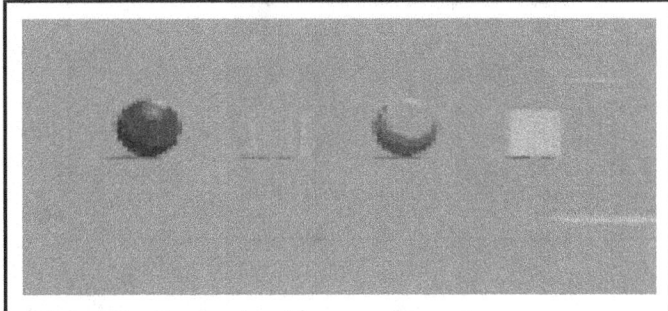

Figure 13c. Synthesized image of targets.

Figure 14. Synthesized configuration on road surface ($R1 - q_0 = 0.1$).

Figure 15. Synthesized configuration on road surface ($R4 - q_0 = 0.1$).

Figure 16. Lighting scheme for crosswalks, Switzerland.

LUMINANCE DESIGN AND PAVEMENT REFLECTION FACTORS

LUMINANCE DESIGN TECHNIQUE

For 25 years, the luminance design technique has been successfully used on major motorways and tunnels in Europe. This method is based on the way the human eye sees; that is, road surfaces are made visible by light reflected from them and entering the eye of the observer.

The panel saw many examples of good lighting that resulted from the use of this design technique. Examples are shown in figures 17, 18, and 19.

European roadways are lit to levels more than twice as high as those in the United States, and with better uniformity. Belgian experts expressed the opinion that a high degree of pavement uniformity yields good driver comfort. They are confident that driver comfort equates to driver safety. They were not, however, aware of any formal studies linking driver comfort to safety.

Based on the Belgian experience, experts suggest that roadways lit to levels between 1 and 2 cd/m^2 produce good visibility, while lighting the roadway to less than 1 cd/m^2 does not yield good visibility. In addition to light level, good visibility in wet conditions also depends on the locations of luminaires. For example, in Finland, the team observed lighting over the roadway.

PAVEMENT REFLECTION FACTORS AND R-TABLES

Because the luminance design method depends on road surfaces being made visible by light reflected from roads and entering the eye of the observer, the reflection properties of pavement become an integral part of the lighting-design process. The existing pavement reflection tables, the R-tables, were published in 1976 and have been used in luminance design

Figure 17. Wevelgem Tunnel, Belgium.

Figure 18. Highway near Helsinki Airport, Finland.

Figure 19. Highway near Helsinki, Finland.

worldwide ever since. The R-tables refer to pavement reflection characteristics under dry road-surface conditions only.

The R-tables are based on two pavement properties: S1, the specularity or pavement shininess; and Qo, the lightness or degree of grayness, from white to black, of a road's surface.

The range of the S1 value determines the class in which pavement is assigned, R1 through R4, as shown in table 3.

TABLE 3. R-TABLE VALUES, BY PAVEMENT CLASS.

Pavement Class	Standard S1	S1 Range
R1	0.25	< 0.42
R2	0.58	>0.42 but < 0.85
R3	1.11	>0.85 but ≤1.35
R4	1.55	>1.35

For accuracy, the average luminance coefficient, Qo, must be determined for the particular pavement under consideration. Typically, the values for Qo are R1 = 0.1, R2 and R3 = 0.7, and R4 = 0.8. However, these typical numbers do vary.

In Belgium, the most commonly encountered road pavements were bituminous asphalts (R3, with Qo from 0.07 to 0.10 cd/m²/lux) and porous asphalts (R2, with Qo from 0.05 to 0.08 cd/ m²/lux). French experts use the real R-value of the roadway, if possible. For quick estimating purposes, however, the following luminance/illuminance conversions for roadway lighting are used in France:

1 cd/m² is produced by 8 lux on light-colored pavement.

1 cd/m² is produced by 18 lux on dark-colored pavement.

1 cd/m² is produced by 14 lux on average-colored pavement.

Swiss, French, and Belgian experts mentioned that a refined analysis of pavement properties is conducted for major projects. The analysis requires that a sample of the future road pavement be measured in the laboratory and the matrix of the reduced reflection coefficient be incorporated into the specifications, along with the minimum required lighting levels and uniformities for the project.

To obtain realistic R-values when evaluating an actual road surface, the Belgians evaluate several core samples and average the results. Outliers are discarded. Furthermore, results of studies have shown that, in the case of porous asphalts, it typically takes between 6 months and a year for the pavement to stabilize in order to obtain reliable R-values.

Figure 20 illustrates why field measurements should be delayed until after the pavement has stabilized. Note that the left lane is not traveled on and is in nearly "as poured" condition, while the right lane shows the typical change in reflection properties caused by traffic. Note that the wheel-rut paths also have a much different specularity than the other pavement areas, which makes it difficult to measure the luminance of the overall pavement. All measured luminance values must be qualified as to the location on the pavement, but methods for determining the overall luminance value from collections of individual points has not been established in Europe or in the United States. This example illustrates the difficulty typically encountered when attempting to enforce luminance specifications or when verifying designs.

Figure 20. Milchbuck Tunnel, Switzerland.

Luminance measurements taken on the two lanes show the right lane, at 140 cd/m², to have twice the luminance level as the one on the left, 70 cd/m².

The Swiss noted problems with standard R-tables and have obtained different results initially than those designed with standard R-tables. In an additional conversation with Werner Riemenschneider, however, he clarified that after the pavement had aged for 6 to 12 months, the Swiss typically found that the measured average values were within 15 percent of the average design value, usually on the high side.

In the good cases the Belgians noted discrepancies of less than 10 percent, when comparing measured luminance levels against calculated levels for pavements, where the reflection characteristics have been determined.

NEW TYPES OF PAVEMENT

As mentioned earlier, pavement types have been invented since the original R-tables were conceived. The French noted increased usage of new surfaces over the past 10 years. These surfaces include a number of wearing courses and porous asphalt, i.e., water-draining pavement.

Porous asphalt stabilizes in a unique way. It becomes more diffuse and its brightness increases, as is shown in figures 21 and 22.

Figures 21 and 22 show computer-generated images of the reflection characteristics when viewed from typical angles down the roadway. The angle most frequently encountered in the past

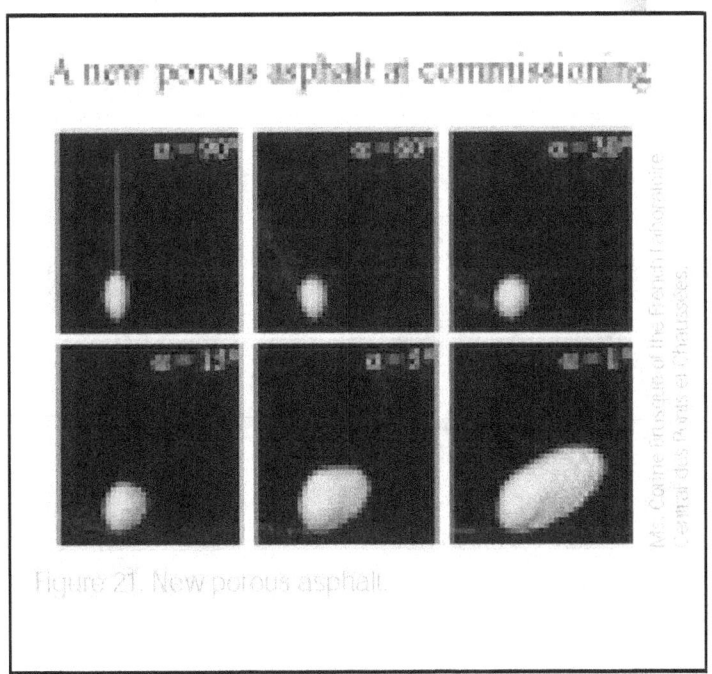
Figure 21. New porous asphalt.

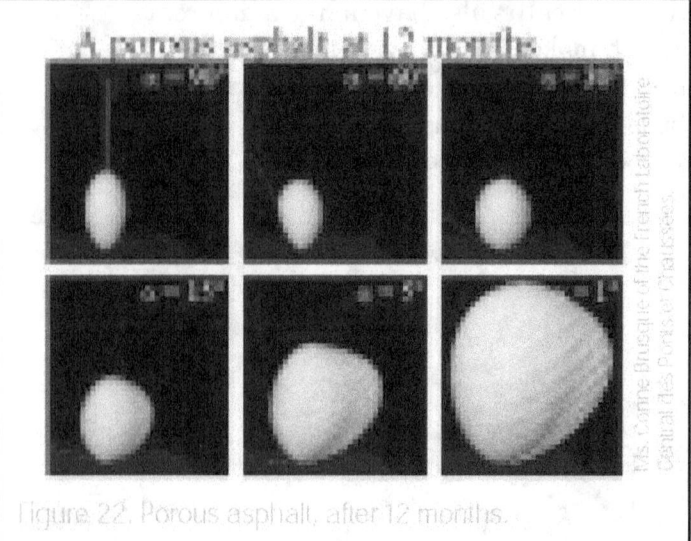

Figure 22. Porous asphalt, after 12 months.

is the 1-degree downward view. The French and the Swiss suggested that additional viewing angles were needed because of lower speeds and urban environments. The angles most frequently mentioned were 3 and 5 degrees.

Appendix D lists a paper by Ms. Corine Brusque that describes how to design lighting for water-draining pavements. The Dutch noted that this type of open-graded asphalt pavement seems to produce better visibility than the older, dense asphalt.

In Switzerland, experts emphasized the importance of dry roadways when conducting field measurements. There are, however, only a couple of summer months during which pavements are dry enough to be measured. In addition, the team heard warnings about dew points and pavement ages. It was noted that, during observations in cold weather (typically October through December), with a clear sky, when conditions were under the dew point, a water film could suddenly appear on the roadway. This film could provide a reflectance differential of 200 percent. Given these difficulties and variations in pavement reflection characteristics, the Swiss typically verify lighting installations with incident light measurements.

PAVEMENT REFLECTION FACTORS: OTHER CONDITIONS

In addition to the R-tables, N-tables are applied in countries that use additional "whiteners" in pavements, which causes the pavement to become very bright. The French noted that, specifically for tunnel lighting, they are researching a special pavement that has white gravel and cream-colored bitumen. Currently, it appears as though the average luminance coefficient, Q_o, changes downward, over a 3-year period. This is still under investigation.

The Scandinavian countries have developed W-tables for use on the wet roadway conditions encountered there (see figure 23).

Figure 23. Wet roadway in Finland (inset, close-up of pavement).

In Finland, the standard R- and W-tables are used, while the other countries that the team visited use only the R-tables. In Switzerland, the W-tables that were developed in Scandinavia are not used. Rather, the Swiss studied 10 typical installations and, based on experience, determined that, for their purposes, 2 cd/m^2 under dry conditions was also adequate in wet conditions. Swiss experts found no operational difficulties with that approach.

PANEL RECOMMENDATION

- Pavement reflectance is an important element of lighting design. The panel recommends that more research, including field measurements, be conducted in order to overcome the acknowledged inadequacy of the R-tables.

TUNNELS

As mentioned earlier, tunnel lighting has been upgraded in the past 10 years. As defined in the technical report, *Guide for the Lighting of Road Tunnels and Underpasses*, CIE 88, 1990, good tunnel lighting should "ensure that traffic, both during day and nighttime, can approach, pass through, and leave a tunnel, at the designated speed, with a degree of safety and comfort not less than that along adjacent stretches of open road."

In the past 10 years, techniques for excavating tunnels have improved, making tunnels a more attractive option. For example, Finland is currently working on a bypass that goes under, not around, the City of Helsinki. Additionally, while typical tunnels are straight drive-throughs, the team observed tunnels in Europe that contained merges and diverges and the team even visited an underground roundabout (figure 24). The underground roundabout, which is part of a mass transit station at Frauenfeld, in the Thurgau Canton, Switzerland, includes not only through roadways, but also an entrance and exit to an underground parking lot (figure 25).

Figure 24. Underground roundabout, Switzerland.

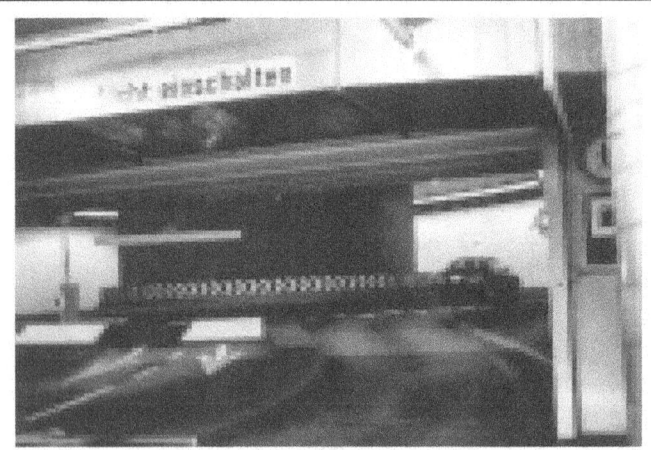
Figure 25. Underground roundabout entrance and exit feeds to underground parking.

Swiss experts believe that good wall luminance is necessary to provide good guidance for motorists. Additionally, they have observed that most people perceive tunnels lit with fluorescent sources to be brighter and more comfortable than tunnels lit on the road to the same level with point sources, probably because of the higher wall luminance normally attained with fluorescent luminaires. Although various light sources are being used, long tunnels are generally lit with electronically ballasted, dimmable fluorescents. That approach easily adapts to the integrated control systems used on all the tunnels that the team observed. The integrated systems use a luminance meter to adjust the light level in the threshold zone. Lighting control systems can be integrated into the traffic management systems, and traffic volume can be added to the control elements. (Traffic control centers are discussed later under Future Developments.)

The primary difficulty in tunnel lighting is determining the correct lighting level to be installed in the threshold. In theory, "just enough light" is necessary to meet the requirement cited in CIE 88. Less than "just enough light" causes the traffic to slow down or frequent crashes to occur. Higher lighting than necessary wastes money, both on installed and maintenance costs. Determinations become more complex with the realization that the lighting decision typically has to be made before the tunnel is built. For all of these reasons, the panel was shown more tunnel lighting than any other type of lighting. The team members were greatly impressed with what they saw (figures 26 to 30).

In the pilot installation of the Wevelgem Tunnel, the threshold luminance of the counter-beam system (Lth, CBL: 400 cd/m^2) was purposely set equal to the threshold luminance of the symmetric system (Lth, Sym: 400 cd/m^2) in order to evaluate the visibility of the targets. Figure 30 shows how the counter-beam lighting is washed out by the outside natural light. This lack of negative contrast in the beginning of the threshold results from daylight penetration (up to 70 m), reflected light from the walls and pavement, and veiling, caused by the natural brightness present in the atmospheric luminance (L_{ATM}). The Belgians found that, whatever the lighting system, there were always invisibility zones for a target with a fixed reflectance factor. These zones are of variable length and position from one system to another. Belgians do not lower the threshold luminance requirement when counter-beam lighting is used.

Figure 26. Tunnel in Lyon, France.

Figure 27. Tunnel in Helsinki, Finland.

When using counter-beam lighting, the Swiss noted that large trucks traveling through tunnels (figure 31) absorb the light coming from the luminaries, creating a lower light level as well as shadows. Also, black trucks are hard to see. Regardless, the Swiss still believe that counter-beam offers the best solution, except in tunnels with a very high level of truck traffic.

Based on experience, the Swiss believe that if sunscreens are used, they must be waterproof. A screen that is not watertight will allow water to drip on the roadway and refreeze.

Figure 28. Tunnel at Schipol Airport, the Netherlands. The sunscreens are not actually needed, because of the counter-beam threshold lighting. They were left in place for aesthetic reasons.

Figure 29. Milchbuck Tunnel, Switzerland.

Figure 30. Wevelgem Tunnel, Belgium.

In the Netherlands, counter-beam lighting, instead of sunscreens, has become more widely used because it is less expensive. Typically, the threshold luminance levels used in the Netherlands, at a design speed of 120 km/h, are 200 to 250 cd/m^2, based on using counter-beam (symmetrical is higher). Dutch designers believe that the current CEN document is 1.5 to 2 times too high and that CIE recommendations are about 20 percent too low.

Recently, the Belgians told us that the Belgian designers have shown, through a European survey made in the European Working Group for Tunnel Lighting (CEN/WG6) that, for the large majority of the tunnels, countries rigorously follow the CIE 88 (1990) recommendations for the threshold zone lighting. Their experience with these lighting levels has been reported positively.

In addition, the Dutch have developed a design method called the "Black Window." The Black Window is used for deciding whether lighting is required for short tunnels. Figure 32 is a diagram of the method, and figure 33 illustrates actual examples.

In the diagram, A, B, C, and D define the area of the entrance portal. E, F, G, and H define the exit.

For D < 20%, lighting is installed.

For D > 50%, no lighting is required.

For D > 20% or < 50%, a study is required.

The method examines what percentage of the typical automobile that passes through the tunnel is visible. If it is 30 percent visible, then no lighting is installed. If it is less than 30 percent visible, then lighting is installed. Lighting is not required on either of the examples in figure 33, because more than 30 percent of a typical automobile is visible.

Based on information from Japan, where tunnel walls are painted a dark color, the Dutch conducted an interesting experiment on the benefits of dark or light walls in tunnels. They eliminated cleaning on one of the tunnels for a period of one and a half years and had no change in crash rates.

Figure 31. Milchbuck Tunnel, Switzerland.

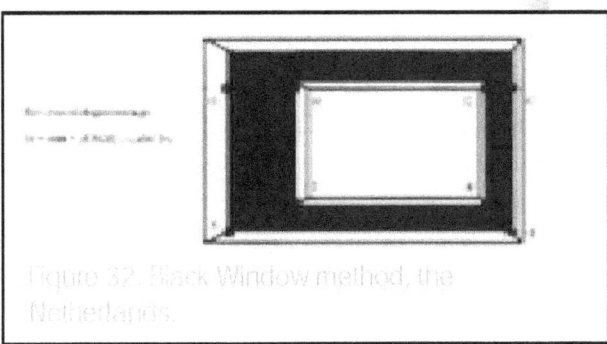

Figure 32. Black Window method, the Netherlands.

Figure 33. Examples of Black Windows.

COUNTER-BEAM AND PRO-BEAM LIGHTING

Today, tunnels are lit to high enough levels to ensure that traffic, both during day and nighttime, can approach, pass through, and leave a tunnel, at the designated speed, with a degree of safety and comfort not less than that along adjacent stretches of open road.

The luminaires that are used to accomplish this typically have one of three types of distribution and effect on objects. The distribution effects are shown in Figures 34 a, b, and c.

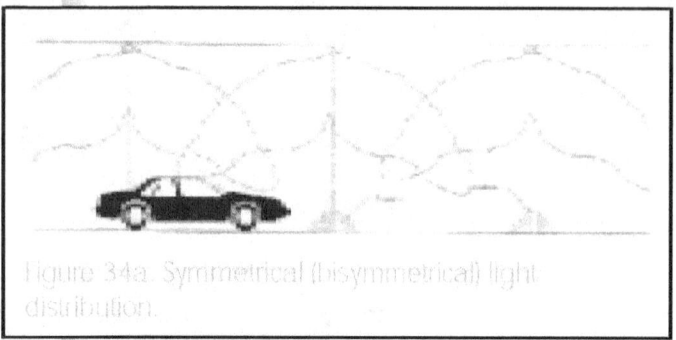

Figure 34a. Symmetrical (bisymmetrical) light distribution.

In figure 34a, light is symmetrically distributed, particularly when linear sources are used. Although a uniform luminance is produced throughout the tunnel, relatively low contrast values are generated.

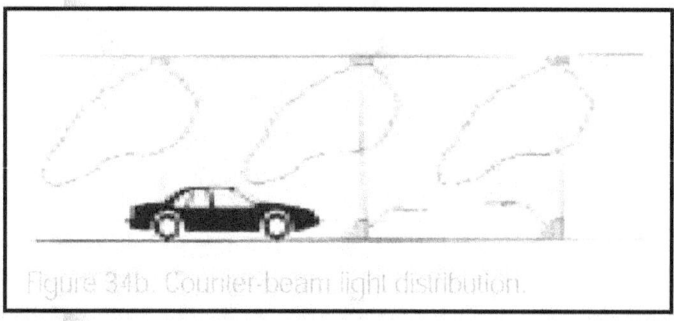

Figure 34b. Counter-beam light distribution.

In figure 34b, light is asymmetrically distributed, with the strongest part of the beam directed toward the approaching driver. This type of lighting provides high pavement luminance and low object luminance, creating negative contrast.

In figure 34c, light is asymmetrically distributed, with the strongest part of the beam directed away from the approaching driver, in the direction of traffic flow. This type of lighting provides high object luminance and low pavement luminance, creating positive contrast.

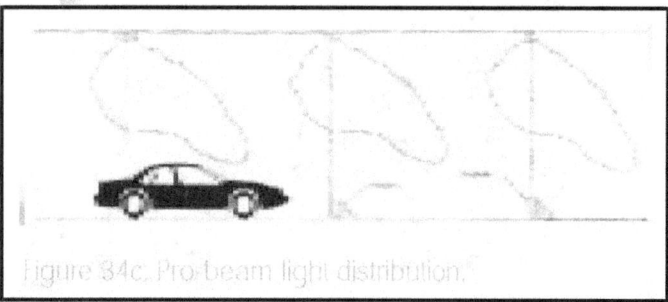

Figure 34c. Pro-beam light distribution.

The panel was most interested in learning about Europe's experience with the different distributions. The French experts suggested that targets disappear under pro-beam and do not use it. In fact, while all of the countries visited have experimented with using pro-beam for threshold lighting in tunnels, none of them use it.

The Swiss have evaluated counter-beam, pro-beam, and symmetric lighting systems. Field measurements and lighting calculations have indicated that, if counter-beam yields a 100 percent light level, then symmetric with the same lumen output yields 70 percent, and pro-beam with the same lumen output yields about 30 percent. This is the case on the mostly used asphalt concrete road surfaces of type R3. On type R1 (less specular) the gain in yield is smaller. The Swiss discovered some problems with counter-beam installations where there is a lot of large truck traffic. Counter-beam lighting is preferred in Switzerland.

Until now, the Belgians have only used symmetrical lighting. They have, however, conducted extensive experiments and found that the best angle (with the vertical) for the main beam in a counter-beam system is 56 degrees.

The Dutch use counter-beam lighting because it is more cost-effective than sunscreens.

HIGH-MAST, DECORATIVE, AND SIGN LIGHTING

HIGH-MAST LIGHTING

The term high-mast lighting generally refers to a group of luminaires mounted at a height of 20 m or more. Its use in interchange lighting leaves the area free of poles and provides motorists with an uncluttered view of the interchange. With careful pole placement, glare is much less of a problem than in conventional lighting. In addition, maintenance can usually be done without disturbing the traffic flow.

In Finland, the panel was shown a nearly completed interchange that was lit with asymmetric floods mounted on high-mast poles. The lighting on that interchange was installed before construction was completed, thereby making it possible to continue work on the interchange after dark (figure 35).

Belgium has several existing interchanges using poles up to 35 m in height (figure 36), but does not have any new high-mast installations. Existing interchanges use either HPS sources, in 400 to 1,000 watt, or 131- to 180-watt LPS, with an asymmetrical photometric distribution.

Figure 35. High-mast lighting, Finland (inset, close-up of luminaire).

Other than in Finland and Belgium, the panel did not observe any new high-mast installations. The French indicated that high-mast lighting was used quite a lot in the 1970s and '80s, but is not currently used.

DECORATIVE LIGHTING

The illuminance design technique is used to light residential, small town, intersection, and conflict areas. The new urban lighting trend in Europe is the use of indirect lighting. Several examples of decorative lighting are shown in figures 37 through 46.

SIGN LIGHTING

In Finland and Switzerland, overhead, directional signs are currently lit with top-mounted luminaires. To save money, however, the Finns are moving away from sign lighting by using micro-prismatic sheeting material. In France, engineering-

Figure 36. High-mast lighting, Belgium.

Figure 37. Parking lot lighting, Helsinki, Finland.

Figure 38. Parking lot lighting, Helsinki Airport, Finland.

Figure 39. To aid recognition, vertical and semispherical illuminance is used in pedestrian areas.

Figures 40a, b, & c. Examples of typical decorative lighting in Zurich, Switzerland.

Figures 41. & 42. Downtown Helsinki, Finland.

grade, retroreflective material has been used, and the French also are moving away from fixed sign lighting.

HIGH-MAST, DECORATIVE, AND SIGN LIGHTING

Figures 43. & 44. Decorative lighting in Finland (inset, detail of luminaire).

Figures 45. & 46. Zurich, Switzerland, at night.

PANEL RECOMMENDATION

- The panel recommends the use of micro-prismatic sheeting materials for unlighted signs mounted overhead and on the left shoulder.

Figure 47. Micro-prismatic sheeting material for signs in Finland.

ROUNDABOUTS

During the tour, the panel observed many new or relatively new roundabouts in Europe. Team members were told that civil and traffic engineers prefer to use roundabouts instead of traffic signals for reasons of safety and efficiency. In addition, in some places, roundabouts are popular with local communities, thus generating demand for more new intersections of that type. Figure 48 shows a roundabout in suburban Paris.

The panel visited the Philips Lighting facility in France. Representatives of Philips indicated that the goal of roundabout lighting was to provide a total view of the roundabout geometry from three levels: long distance, nearby (100 m), and at the entrance. Figure 49 shows a roundabout at the Philips facility.

Figure 48. Roundabout, Paris, France.

ROUNDABOUT CATEGORIES

The French divide roundabouts into the following three categories:

Urban location	Systematically lit
Suburban location	Recommend lighting
Rural location	Not lighted, unless there is already lighting in the vicinity

LUMINAIRE LOCATIONS

In France, outer edge or central island luminaire locations are permitted. In Finland, Belgium, and Switzerland, roundabouts are lit from the outer edge.

Swiss and Belgian designers initially lit from the center, but found poor results with negative contrast in crosswalks and roundabouts. The Swiss design specifically addresses pedestrian crosswalks and roundabouts by providing positive contrast.

Figure 49. Roundabout at Philips Outdoor Lighting Application Center, La Valbonne, France.

Swiss and Belgian experts believe that proper placement of luminaires (figure 50) will provide positive contrast on pedestrians and automobiles, thereby improving recognition. In addition, highly visible (retroreflective) materials are used extensively on the curb (or periphery) of the center of the island to enhance its conspicuity.

ROUNDABOUT LIGHT LEVELS

In Finland, designers strive to have 30 percent more light on the roundabout than on the approaching roads. Swiss roundabouts are designed for homogeneous lighting at a level 50 percent above the best approach road.

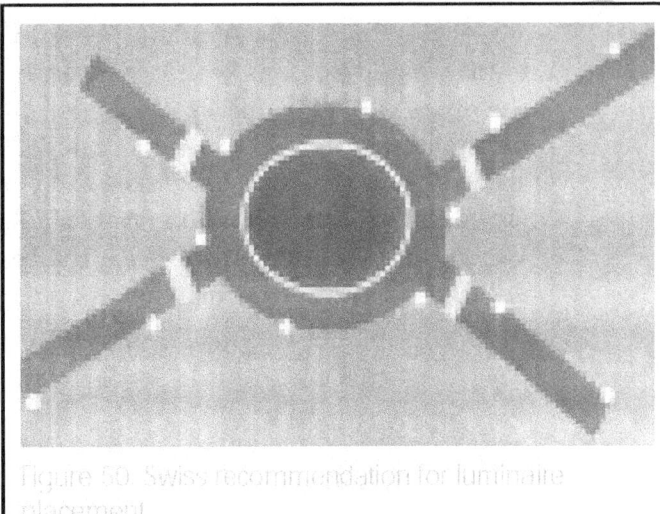

Figure 50: Swiss recommendation for luminaire placement.

In Belgium, roundabout design is for a 40-lux horizontal illuminance light level with a uniformity greater than 40 percent. In addition, Belgian experts have found that providing a 15-lux vertical level at 2 m from the outside edge of the central island improves the general perception of the roundabout. The central island has to be materialized by vertical (retroreflective) elements on the central island with a minimal frontal surface. They believe that the entrances of the roundabout (120 to 150 m before the roundabout) also have to be lighted.

PANEL RECOMMENDATION

- The panel recommends that the European experience in roundabout lighting be synthesized and consolidated in the AASHTO *Lighting Guide* for U.S. application.

SAFETY IMPLICATIONS

The Swiss Council for Accident Prevention is a private, politically independent foundation, which has been legally entrusted with the task of preventing accidents in the areas of road traffic, sports, home, and leisure. Tables 4 and 5 include comparative information given to the team members by Mr. Paul Reichardt of the Swiss Council. Table 4 gives the impression that the United States has a very serious problem. However, when the raw data are normalized with millions of km driven (table 5), the United States is comparable with the best of the countries cited. The United States and the countries that the team visited are highlighted in the tables. The data are from the International Road Traffic and Accident Database (IRTAD).

TABLE 4. FATALITIES IN ROAD ACCIDENTS.

COUNTRY	1994	1995	1996	1997	1998
United States	40,716	41,798	42,065	41,967	-
Turkey	-	-	-	6,735	6,308
Sweden	589	572	537	541	-
Korea	11,600	11,871	14,551	13,343	10,416
Poland	6,744	6,900	6,359	7,310	7,080
Portugal	2,504	2,711	2,730	-	2,425
New Zealand	580	581	514	540	-
Netherlands	1,298	1,334	1,180	1,163	1,066
Norway	283	305	255	303	352
Luxembourg	74	-	-	60	57
Japan	12,768	12,670	11,674	11,254	10,805
Iceland	12	24	10	15	27
Ireland	404	437	453	472	-
Italy	7,104	7,033	6,688	6,724	-
Hungary	1,562	1,589	1,370	1,391	1,371
Greece	2,253	2,411	2,063	2,199	-
Great Britain	3,650	3,621	3,598	3,599	-
Finland	480	441	404	438	-
France	9,019	8,891	8,541	8,444	8,918
Spain	5,615	5,751	5,483	5,604	-
Denmark	546	582	514	489	-
Germany	9,814	9,454	8,758	8,549	7,776
Czech Republic	1,637	1,588	1,568	1,597	1,360
Switzerland	679	692	616	587	597
Canada	3,263	3,347	3,092	3,064	-
Belgium	1,692	1,449	1,356	1,364	-
Australia	1,938	2,013	1,970	1,767	1,763
Austria	1,338	1,210	1,027	1,105	963

TABLE 5. FATALITIES PER 1 BILLION VEHICLE KM TRAVELED (VKT).

COUNTRY	1994	1995	1996	1997	1998
United States	11	11	11	10	-
Turkey	-	-	-	139	-
Sweden	-	-	-	-	-
Korea	-	-	-	-	-
Poland	-	-	-	-	-
Portugal	-	-	-	-	-
New Zealand	-	-	-	-	-
Netherlands	12	12	11	-	-
Norway	10	10	8	10	-
Luxembourg	-	-	-	-	-
Japan	18	18	16	15	-
Iceland	6	13	5	8	14
Ireland	14	14	13	-	-
Italy	-	-	-	-	-
Hungary	-	-	-	-	-
Greece	34	35	28	29	-
Great Britain	9	8	8	8	-
Finland	12	10	10	10	-
France	19	18	17	16	-
Spain	-	-	-	-	-
Denmark	14	14	12	-	-
Germany	17	16	14	14	-
Czech Republic	58	49	-	-	-
Switzerland	14	14	12	11	-
Canada	-	-	-	-	-
Belgium	21	18	17	16	-
Australia	-	12	-	-	-
Austria	21	19	15	16	-

At night, a human's visual capabilities are impaired, and visibility is reduced. Road crashes at night are disproportionately high in numbers and severity when compared with the daytime. In the United States, while only 25 percent of the travel occurs during nighttime, about 55 percent of the fatal crashes occur after sunset. Weighted for km traveled, the nighttime fatality rate is three times the daytime figure.[5] The major factor contributing to this problem is darkness, because of its influence on a driver's behavior and ability. Thus, logically, road lighting is a potential countermeasure.

Most of the countries reported significant safety benefits in term of crashes, injuries, and fatalities when road lighting was installed. Some sample statistics follow:

- Finland reported 20 to 30 percent reductions.
- A Norwegian study that was cited revealed a 65 percent reduction in nighttime fatalities, a 30 percent reduction in injuries, and a 15 percent reduction in property damage.
- Dutch studies showed reductions of 18 to 23 percent.

In Finland, traffic fatalities were lowered from 1,000 in 1971 to 410 in 1998. Because there are long periods of darkness during the year, it is likely that roadway lighting can be credited for some portion of the decrease in fatalities.

Swiss representatives reported that crash rates are lower in appropriately lighted tunnels than on other roadways.

Arguably the best data on this subject are available in the technical report, *Road Lighting as an Accident Countermeasure*, CIE 93, 1992. The report includes rigorous analysis of 62 lighting and crash studies from 15 countries. Eighty-five percent of the results show that lighting was beneficial, with about one-third of these studies having statistical significance.

These data lead to the general conclusion that road lighting on traffic routes will reduce the incidence of nighttime accidents. Depending on the class of road and the accident classification involved, the statistically significant results show reductions of between 13 and 75 percent. Some of the specifics are:

Figure 51. The Swiss "Vision Zero" program.

	Percent Reduction
Motorways and semi-motorways	20
Other roads for motorized traffic	25
All-purpose roadways	30

The Swiss have launched an ambitious program known as "Vision Zero" (figure 51). Its purpose is to improve roadways such that there are "no fatalities in traffic accidents." The graphs in figure 52 show the number of road accidents from 1945 to 1995 as compared with the number of vehicles on the road during the same period.

OTHER OBSERVATIONS

Finnish representatives referred to an interesting experiment that was conducted in southern Finland. The road lighting was reduced from 1.5 cd/m^2 to no lighting at all. The result was a 25 percent increase in the accident rate. When the lighting was reduced from 1.5 cd/m^2 to 0.75 cd/m^2, the accident rate increased 13 percent.

The Finnish Road Administration offers incentives to road district personnel who implement creative safety improvements. A monetary reward is provided if accident rates are reduced.

Based on the Dutch experience with reduced lighting levels during the energy crisis, light levels on motorways have been reduced to the range used in the United States, apparently without a noticeable increase in accident rates.

Dutch designers installed an experimental, dynamically lit roadway that can be operated at three lighting levels, depending on the amount of traffic and weather conditions. The normal level is 1 cd/m², the high level is 2 cd/m², and the low level is 0.2 cd/m². Experts were unable to detect statistical differences in accidents between 1 cd/m² and 2 cd/m²; however, the sample size was very small. Accident rates for the 0.2-cd/m² system, when it is operated at low traffic volumes, have been acceptable, and a second system that only operates at 1 cd/m² and 0.2 cd/m² has been installed.

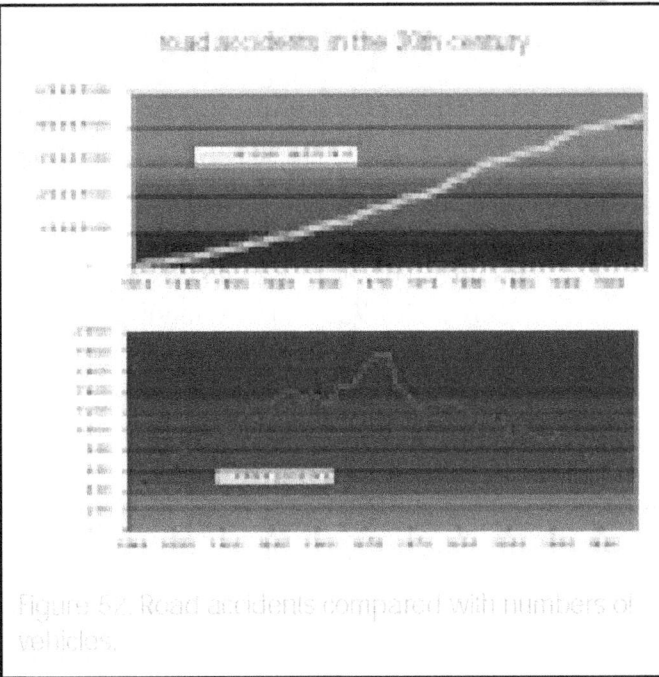

Figure 52. Road accidents compared with numbers of vehicles.

In Switzerland, the Zurich police provided the panel with an extensive investigative report on accidents in the Gubrist Tunnel, where there have been 75 accidents over a 30-month period. The investigation included an analysis of the lighting in the tunnel and videotape of a number of accidents. The panel found it highly interesting that the police were analyzing the causes of crashes.

PANEL RECOMMENDATION

- The panel recommends the development of a uniform (State-to-State) accident reporting system that includes more accurate descriptions of the lighting conditions at crash scenes.

FUTURE DEVELOPMENTS

In the area of future developments, the panel was very interested in investigating cutting-edge lighting research as well as anticipated major advancements in the art and science of roadway lighting.

NEW EUROPEAN STANDARDS

All the countries that the team visited, with the exception of Switzerland, are members of the EU, and substantial effort is going into the harmonized CEN Lighting Standards. When the harmonized CEN documents are adopted, they will replace individual countries' standards, which were generally based on the CIE. This is an example of the impact of the EU now and in the future.

TRAFFIC CONTROL CENTERS

The panel visited two traffic control centers (TCC): one in Finland and one in Switzerland (figures 53 and 54).

The TCCs are used to improve traffic flow, provide traffic information, and control and manage traffic demand. The Finnish center can monitor the weather throughout the entire country and, when needed, give information directly to motorists by interrupting car radio programs. Motorists do not need to be tuned to any particular frequency. The center can also remotely change posted speed limits.

Figures 53 & 54. Views of a TCC in Switzerland (left) and Finland (right).

DYNAMIC ROADWAY LIGHTING

In the Netherlands, the origins of dynamic roadway lighting can be traced to the Energy Crisis of the 1970s. During that period, some luminaires were turned off to save energy. While there was an increase in accidents, it was not a large increase. (Some areas of the United States noted significant increases in accidents when the same approach was used). Over the following 15 years, there was movement by the Netherlands to lower the lighting levels from 2 cd/m^2 (as recommended by CIE) to 1 cd/m^2, retaining the recommended uniformity ratios.

Figure 55. Low level of roadway lighting, the Netherlands.

Since 1995, the Netherlands has installed and operated a dynamically lighted roadway that can be adjusted to any of three lighting levels, depending on the amount of traffic, time of day, and weather conditions. The low level is 0.2 cd/m^2 (figure 55), the normal level is 1 cd/m^2 (figure 56), and the high level is 2.0 cd/m^2 (figure 57). The different light levels are obtained through the use of electronically controlled, dimmable HPS ballasts.

Figure 56. Normal level of roadway lighting.

To set a baseline for the dynamic road section, Dutch experts have collected and analyzed accident data. Unfortunately, the dynamic section was too short and the statistical sample size was too small to draw conclusions between the 1-cd/m^2 and 2-cd/m^2 light levels. In an evaluation of an extensive set of methods (inductive loop detectors, instrumented vehicles, video observations, questionnaires), it was concluded that, under low traffic volumes (less than 800 vehicles per hour) and favorable weather conditions, the low level (0.2 cd/m^2) can be applied. Accidents rates

Figure 57. High level of roadway lighting.

for the low-level lighting have been acceptable. To continue gathering information on dynamic road lighting, the Dutch have installed a second system, which only operates at 1 cd/m^2 and 0.2 cd/m^2.

In Finland, a consortium of three organizations (FORTUM, SITO, and VTT) is experimenting with a dynamic road lighting system on a 3.5-km segment of Route 1

(Oinola to Saukkola), with about 9,000 ADT, on a two-lane road. The system uses a continuous integration of traffic volume and weather conditions to determine the speed limits and roadway lighting levels. A measuring device (meter) is used to determine whether the pavement is wet, dry, or snow-covered. The control system tries to keep the luminance of the roadway constant by varying the lumen output of each luminaire. The meters also can determine which luminaires are not functioning properly. Dimmers made by Philips Telemanagement control individual luminaires. The schedule called for testing of the system to begin in autumn of 2000.

GUIDANCE SYSTEMS

In the mid-1990s, environmental studies concluded that a lighted roadway could be a barrier to wildlife movement. In addition, a number of environmentalists suggested that darkness was a natural and good thing. A number of environmentally sensitive areas in the north of the Netherlands are referred to as "scenic areas." In the scenic areas, the current lighting approach is multifaceted and includes not installing lighting, installing lighting that can be dimmed, and an active investigation into the use of lighting as a guidance system.

Experts in the Netherlands are researching the acceptability of a number of different types of guidance systems. Under investigation are light-emitting diode (LED) pavement markers, LED post delineators, LED pavement-marker stripes, fiber-optic "side sights" (fiber optics attached to a guardrail, with light emanating along the entire length), and fiber optic "end lights" (in-pavement fiber optics with ends extending up and out of the pavement surface at fixed intervals with light emanating from the tips of the cables). These systems are used where additional guidance is needed and are typically operated between 11:30 PM and 6 AM. Figure 58 illustrates the types of systems.

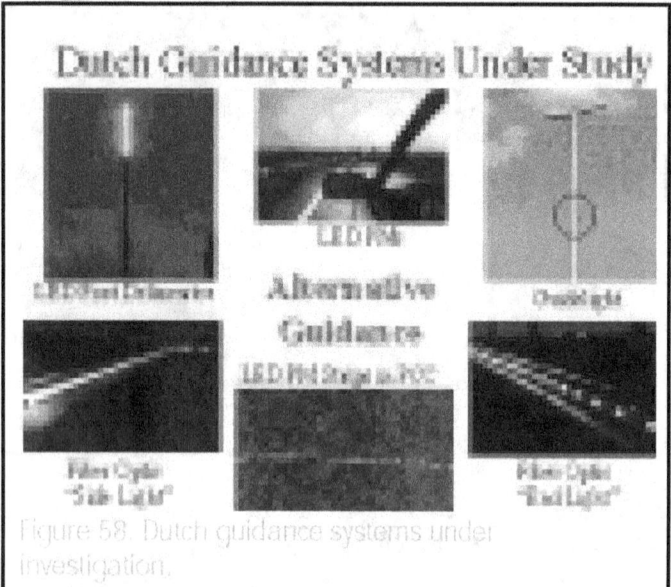

Figure 58. Dutch guidance systems under investigation.

To date, findings of the Dutch investigations indicate the following:

1. In-road systems seem to give the best guidance information to motorists.

2. Guardrail-mounted systems do not always relate to the roadway.

3. In-road LEDs should match the lane line stripe color.

4. Solar-powered LEDs typically take 14 hours to run down and appear to have approximately a 5- to 5.5-year service life.

5. Guardrail-mounted equipment requires costly repair when a car crashes into it.

6. In-pavement systems pose a challenge during resurfacing operations.

The panel observed in-road, fiber-optic delineators in Switzerland, as well, which are shown in figures 59a and b.

In addition to the research by the Dutch, the French have a study under way comparing lighting, retroreflectivity, and active luminous devices. Also, Helsinki University is working in the area of mesopic vision (luminance levels that are typically used in roadway lighting) and use of LEDs in lighting. There is an extensive amount of research being conducted in this area.

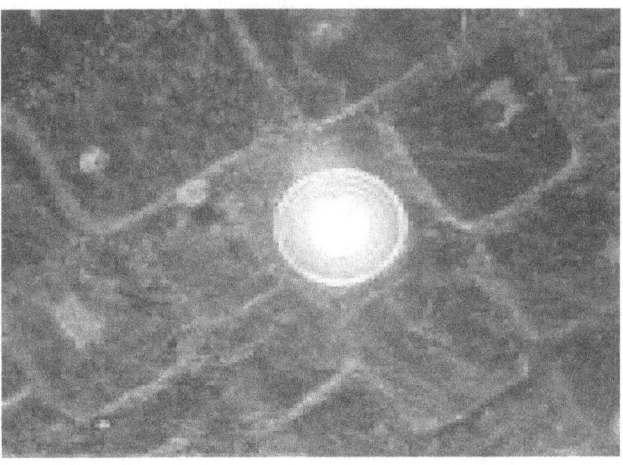

Figures 59a & 59b. In-road, fiber-optic delineators.

PAVEMENT REFLECTION QUALITIES

Some of the newer surfaces that are not included in the development of the original R-tables include quiet and water-draining pavements, as well as very thin, asphaltic concretes and surface dressings. Additionally, there is an increase in the use of bright and colored road surfaces, as shown in figure 60.

Because of the evolution of road surface technology, the French are conducting research in the area of photometric properties of road surfaces. Figures 61 and 62 show applications of colored pavements.

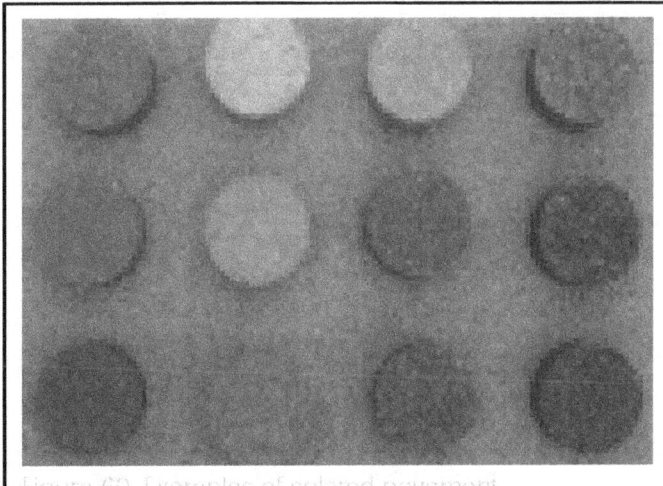

Figure 60. Examples of colored pavement.

There is a need for new pavement reflectance measuring equipment as well as data for observation angles, other than 1 degree downward. Figure 63 illustrates a typical view that a motorist sees inside a tunnel. Here, the driver will usually shorten his gaze to closer objects both on and off the roadway, depending on his rate of speed. Data are needed for the closer observation angles.

In Belgium, R-Tech is building a reflectometer to measure pavement reflectance at varying alpha, beta, and gamma angles. In France, with use of a ray tracing technique, the French are developing a virtual reflectometer (figure 64) to predict current and future pavement reflectance for all angles.

TUNNELS

Worldwide, the volume of vehicular traffic is increasing. To keep traffic moving through tunnels in the daytime, the lighting community has increased the amount of light installed so that the "black hole" that used to be present at the tunnel portal has been improved to a "gray hole." The intent is to make sure drivers can see well enough into the tunnel so they don't slow down when entering the tunnel. While this has been accomplished, it has been expensive to install and operate – offsetting sunlight is not cheap! As a result, those responsible for lighting tunnels are always looking for ways to accomplish the necessary visual task for less money.

In Switzerland, an example of partial tunnel lighting for a 120-m-long tunnel was cited. The partial-lighting approach for short tunnels may provide the needed visibility while saving energy. It utilizes a known phenomenon: natural daytime lighting typically penetrates the portals of the tunnel about 40 m. Taking advantage of that, artificial lighting is only installed in the middle 40 m of the tunnel. Installing only one-third as much lighting saves a great deal of energy, which makes it very appealing.

Figure 61. & 62 Application of colored pavement.

Johan Alferdinck of the research firm TNO Human Factors, the Netherlands, presented a paper examining the effects of light sources on color contrast in tunnel lighting. The purpose of the research is to answer the question, "Based on luminance contrast, does the color of the light source add anything to this, so that I can reduce the light level in the threshold?" While further research needs to be conducted, part of the conclusion reached is that the use of colored light sources in tunnel lighting is superior in all conditions. Figure 65 shows colored targets with two

Figure 63. Typical motorist's view of tunnel.

different types of lighting. Note the difficulty in target detection when veiling luminance is added.

RESEARCH NEEDS

Control of light levels in tunnels has traditionally been done using a technique that looks at a 20-degree cone and is referred to as "L20." Another technique, called Lseq., which uses a cone of view that is more heavily weighted in the center, also can be used to control the lighting. Jean-Marie Dijon, along with R-Tech, in Belgium, believe that both the L20 and Lseq. controls should be placed on the same tunnel and comparisons made to determine which method performed the best.

Dr. Peter Blaser, in Switzerland, stated that there is a need for research on the visual task in tunnels. He suggested that, for today's traffic conditions, small targets in an empty tunnel do not adequately describe the situation.

Figure 64. Virtual reflectometer, France.

OVERALL RESEARCH IMPRESSIONS

The active research items discussed above are only a portion of the large number of projects under way. As was clear from visiting five countries, the Europeans take the task of advancing the art and science of lighting seriously. While some of the research is funded by private industry, much of it is paid for by various government agencies. For example, the Finnish Road Administration spends 1.5 percent of its annual budget on research and development.

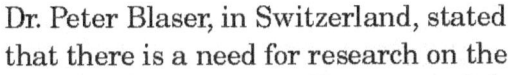

Figure 65. Effects of tunnel lighting color.

PANEL RECOMMENDATIONS

- Investigate the application of the concepts of dimming lighting systems, turning off lighting systems, and alternative guidance systems as approaches to more dynamic management of roadway lighting.
- Encourage innovative experimentation on active roadways and test tracks.
- Further evaluate European standards, practices, and guidance documents to determine applicability in the United States.

SUMMARY OF RESEARCH RECOMMENDATIONS

1. Investigate the use of video-based data acquisition and Fast Fourier Transform analysis for evaluation of the effectiveness (retroreflectivity degradation) of traffic control devices such as sign lettering, pavement markings, and delineators.

2. Evaluate driver information needs at night, considering the following: safe stopping distance, navigational information needs, object in roadway informational needs, visibility needs in periphery (roadside) vision, probabilities of driver's attention being given to the various areas, the change in driver's scan for information habits with and without lighting (including just partial lighting) and also with variations in traffic volume, and the adequacy of small targets describing the overall visibility of the roadway (or providing the needed information).

3. Quantify visibility differences in tunnels between positive- and negative-contrast lighting systems, especially in low tunnel ceilings with heavy truck traffic.

4. Compare lighting systems using Information Theory (IT) (Fast Fourier Transforms) to discern differences between systems designed by the illuminance method, luminance method, and the STV method. Make field evaluations of pavement reflectance and make comparisons in the variation of information based on variations in pavement reflectance.

5. Develop new bidirectional pavement reflectance distribution functions for all pavement types. Investigate variation in reflectivity due to spectral content of lamp. Evaluate Fast Fourier Transforms of pavement texture for correlation to pavement reflectance.

6. Develop measuring techniques and standards for off-roadway glare sources. Research should include the effectiveness of adding or increasing roadway lighting levels to mitigate adverse effects of off-roadway lighting.

7. Investigate the adverse effects of glare on pedestrians and bicyclists sufficient to allow designers to establish limits for such glare. Consider the benefits of the pedestrian's visibility versus the ability of the pedestrian to be seen.

ACKNOWLEDGMENTS

The trip described in this report was successful thanks to the contributions and sacrifices of a large number of individuals. First, and foremost, the panel members thank the engineers and other transportation officials from the five countries visited. These individuals gladly gave their time and resources to make us feel comfortable and to provide the panel with the latest technical information from their respective organizations. The panel met too many individuals to list here, but they are listed in appendix C. In addition to the people listed in the appendix, the panel members thank the behind-the-scenes individuals who worked on the logistical aspects of the trip. In many cases, the panel members may never have met these people, but we recognize their valuable contribution to the success of this trip.

The panel members would never have been able to start and successfully complete the trip without the assistance of the staff from American Trade Initiatives, Inc. (ATI). ATI, under contract with the Federal Highway Administration (FHWA), handled the pre-trip logistics, escorted the delegation throughout the trip, and provided the support needed to prepare and publish this report. The panel members would like to recognize the following staff of ATI:

- Joe Conn, for his guidance and assistance in organizing the trip.
- John O'Neill, for his guidance, counsel, and leadership as our escort during the tour.
- Alexandra Doumani, for her assistance in preparing the report and making the travel arrangements.
- Marie-Dominique Gorrigan, for advance work and translation in Paris.

This trip was made possible with support and funding from the FHWA Midwestern Resource Center and the American Association of State Highway and Transportation Officials (AASHTO). In particular, the panel members thank Donald Symmes and Hana Maier, from the FHWA Office of International Programs, and Dave Hensing and Kyung Ku Lim, of AASHTO, for sponsoring the trip and allowing panel members to observe and report on practices in European countries. Additionally, the panel members thank the staff of the National Coopoerative Highway Research Program (NCHRP) for their assistance.

Finally, the panel members express appreciation to the many individuals who served as interpreters throughout the trip. Although many of our hosts spoke excellent English, having an interpreter who was familiar with the technical terms of the lighting profession was of immense assistance to the panel.

APPENDIX A
PANEL MEMBERS

Karl A. Burkett, P.E. (Co-Chair)
Texas DOT
125 East 11th Street
Austin, TX 78701
Tel: 512-416-3121
Fax: 512-416-3161
Email: kburkett@dot.state.tx.us

Dale Wilken (Co-Chair)
FHWA
10 S. Howard St., Suite 4000
Baltimore, MD 21201
Tel: 410-962-0093
Fax: 410-962-3655
Email: dale.wilken@fhwa.dot.gov

James A. Havard (Report Facilitator)
139 Georgetown Rd.
Hendersonville, NC 28739
Tel: 828-692-1324
Fax: 828-692-0849
Email: jjhavard@bellsouth.net

Balu Ananthanarayanan, P.E.
Wisconsin DOT
Bureau of Highway Operations (Room 501)
4802 Sheboygan Ave.
Madison, WI 53707
Tel: 608-266-0299
Fax: 608-261-6295
Email: balu.ananthanarayanan@dot.state.wi.us

John Arens
FHWA, Turner-Fairbank Highway Research Center
6300 Georgetown Pike
McLean, VA 22101-2296
Tel: 202-493-3364
Fax: 202-493-3415
Email: john.arens@fhwa.dot.gov

Patrick Hasson
FHWA, Midwestern Resource Center
19900 Governors Dr., Suite 301
Olympia Fields, IL 60461
Tel: 708-283-3595
Fax: 708-283-3501
Email: patrick.hasson@fhwa.dot.gov

Paul J. Lutkevich, P.E.
Parsons Brinckerhoff
75 Arlington St.
Boston, MA 02116
Tel: 617-426-7330
Fax: 617-482-8487
Email: lutkevich@pbworld.com

Jeff Unick, P.E.
Pennsylvania DOT
Bureau of Design
PO Box 3060
Harrisburg, PA 17105-3060
Tel: 717-772-3077
Fax: 717-705-2378
Email: jlunick@hotmail.com

C. Paul Watson, P.E.
State Electrical Engineer
Alabama DOT
PO Box 3050
Montgomery, AL 36130-3050
Tel: 334-242-6160
Fax: 334-834-3352
Email: watsonp@dot.state.al.us

BIOGRAPHIC SKETCHES

Karl A. Burkett, P.E. (Panel Co-Chair, AASHTO), is currently the Senior Lighting Design Engineer with the Texas Department of Transportation (TX DOT). Mr. Burkett has been with TX DOT since 1980. He is a licensed Professional Engineer in Texas and a member of the National Society of Professional Engineers as well as the Texas Society of Professional Engineers. Mr. Burkett is Chairman of the AASHTO Task Force for Roadway Lighting and a member of two panels of lighting experts for the National Academy of Sciences Cooperative Highway Research Program research projects. He is also a member of the Illuminating Engineering Society of North America (IESNA), President of the IESNA Texas Capitol Section (Austin, TX), member of IESNA Roadway Lighting Committee and several sub-committees, and Project Director of several TX DOT research projects, including "Evaluation of Roadway Lighting Systems Designed by Small Target Visibility (STV) Methods." Mr. Burkett received the IESNA Roadway Lighting Committee's Roadway Lighting Design Award in 1998. He holds a BSEE degree from the University of Texas at Austin.

Dale E. Wilken (Panel Co-Chair, FHWA) is the Director of the FHWA's Eastern Resource Center located in Baltimore, Maryland. Mr. Wilken directs a staff of technical and program specialists who are responsible for providing technical assistance, training, and deployment of new technology relating to roadway and structure planning, design, and construction to FHWA Division (State) offices and State Departments of Transportation in 15 mid-Atlantic and Northeastern states. Mr. Wilken and his staff also provide these services to metropolitan planning organizations, local transportation agencies, and other customers in these States. In addition, Mr. Wilken supervises the Division Administrators who head the 15 mid-Atlantic and Northeastern Division Offices. Mr. Wilken has previously served as Regional Administrator in Chicago and Division Administrator in Salem, Oregon, as well as Chief of the Environmental Review Branch, FHWA Headquarters, in Washington, D.C. While serving in the Montana Division, his responsibilities included review of highway and structural lighting plans and proposals. Mr. Wilken is a graduate of Bradley University with a Bachelor of Science in Civil Engineering. He is a member of the American Society of Civil Engineers. (Mr. Wilken retired from the FHWA in January 2001.)

Balu Ananthanarayanan, P.E., is the State Electrical Engineer for the Wisconsin Department of Transportation in Madison, Wisconsin. He is currently responsible for the development of Electrical Engineering Policy and Illumination standards for the design, operation, and maintenance of all Wisconsin DOT electrical installations. Mr. Ananthanarayanan has over 26 years of experience as a practicing Electrical Engineer, and he is a member of several national transportation organizations. He is the Wisconsin DOT's representative to the AASHTO Task Force on Roadway Lighting and also the Task Force's secretary. Mr. Ananthanarayanan has been an active participant on several NCHRP project panels as well as on TRB panels and is a past Chair of the Roadway Lighting Committee of the IESNA. He is a graduate of Arizona State University, with an Electrical Engineering degree, and is a licensed Professional Engineer in Wisconsin and Arizona.

John Arens is the manager of the Photometric and Visibility Laboratory of the FHWA Turner-Fairbank Highway Research Center in McLean, Virginia. Mr. Arens conducts and is responsible for photometric and colorimetric material evaluations used for signing and marking of roadways. He is also responsible for lighting criteria applicable to roads, tunnels, signs, and rest areas, as they pertain to federally financed roadways. Mr. Arens previously worked for the Lighting Division of Westinghouse Electric Corporation for 22 years as a test engineer, design engineer, and marketing engineer and for FHWA for 25 years as a lighting engineer and in lighting/visibility-related research. He has been a member of the IESNA since 1958, and he has been active on the Roadway Lighting, the Testing Procedures, and the Papers Committees. He is a past President of both the Capitol Section (Washington D.C.) and the Cleveland Section. He is a member of the International Commission on Illumination (CIE), Divisions 2 and 4, the Council for Optical and Radiation Measurements (CORM), the Visibility Committee of the Transportation Research Council, and the AASHTO Task Force responsible for rewriting *An Informational Guide for Roadway Lighting*. Mr. Arens is a graduate of Cleveland State University with a degree in Electrical Engineering. (Mr. Arens retired from the FHWA in 2000.)

Patrick Hasson is the Safety Team Leader in the FHWA Midwestern Resource Center in Olympia Fields, Illinois. Mr. Hasson provides safety-related technical assistance and advice to FHWA Division Offices, State DOTs, and other transport organizations and officials in the 10 Midwestern States. He also manages and/or provides technical support for a variety of regional, national, and international programs and activities. He is currently the coordinator for the national Stop Red Light Running Program. Mr. Hasson spent 2 years in the Road Transport Research Program at the Organization for Economic Cooperation and Development in Paris, France, where he was involved in a variety of international research projects focused on safety, infrastructure, and transport operations. He is a member of the Institute for Transportation Engineers (ITE) and participates in safety committees with the ITE and the Transportation Research Board. Mr. Hasson holds a BS in Engineering from the University of Maryland and an MS in Engineering from Cornell University.

Jim Havard (Report Facilitator) is one of the principals in LITES, a lighting and information consulting business. He has been involved in all phases of the illumination industry for more than 35 years. His work has included product conception, design, manufacturing, marketing, and application engineering. Mr. Havard is active on the ANSI C136 Committee, where he is the retiring Committee Secretary, and on the IESNA Roadway Lighting Committee, where he is the current Committee Secretary. Mr. Havard is also a member of the Sign Lighting, Tunnel and Standard Practice Subcommittees, and he is the CIE Div. 4 Deputy Representative and is a member of the Visibility Design for Roadway Lighting, and Tunnel Lighting Technical Committees. Mr. Havard is the USNC-IEC, retiring Deputy Technical Representative and Past-President of the IESNA, Street & Area Lighting Committee. He was the Chairman of the Standard Practice Subcommittee that authored the latest revision to RP-8 *American National Standard Practice for Roadway Lighting* and was also a member of the Visibility Taskforce that researched visibility on U.S. roadways.

Paul J. Lutkevich, P.E., is currently the Chair of the Illuminating Engineering Society's Tunnel Lighting Sub-Committee as well as the incoming Chair for the Roadway Lighting Committee. He is also an active member of the Commission Internationale De L'Eclairage (International Committee on Illumination). Mr. Lutkevich is a Senior Supervising Engineer for Parsons Brinckerhoff in Boston, Massachusetts. Over the past 18 years, Mr. Lutkevich has been involved in the design of more than $200 million of lighting systems for roads and tunnels throughout North America. This includes Boston's Central Artery/Tunnel Project, which consists of 161 lane miles of roadway, half of which is in tunnels. He has received awards for his work, including an International Illuminating Design Award for the lighting modernization of Boston's Callahan Tunnel. That lighting system, the first of its kind, is the world's largest light-guide installation. He is a licensed Professional Engineer in several States. Mr. Lutkevich is a graduate of the University of Massachusetts and holds a Bachelor's degree in Electrical Engineering Technology.

Jeff Unick, P.E., is a highway lighting designer for the Pennsylvania Department of Transportation in Harrisburg, Pennsylvania. Mr. Unick is Chief of the Highway Lighting Section in the Bureau of Design. His section provides all of the design, construction, and maintenance expertise for roadway lighting in the DOT. Typical lighting projects include roadway interchanges, tunnels, bridges, roadside rest areas, park and ride areas, and pedestrian ways. He is a member of the AASHTO Task Force on Roadway Lighting and a member of the IESNA. Mr. Unick has a degree in

Electrical Engineering from Pennsylvania State University and is a licensed Professional Engineer in Pennsylvania.

C. Paul Watson, P.E., is the State Electrical Engineer in the Design Bureau of the Alabama DOT (ALDOT) in Montgomery, Alabama. Mr. Watson currently directs the preparation of plans for roadway, bridge and tunnel lighting, traffic signals, and Automated Traffic Management Systems (ATMS). His current projects include a $40 million ATMS for the Jefferson/Shelby County Metro area, lighting of a high-speed fly-over ramp interchange on I-459, and several coordinated traffic signal system projects. Prior to his current assignment, he was instrumental in development of a lighting pole foundation and wiring system to meet the requirements of the AASHTO *Roadside Design Guide*. He is a licensed Professional Engineer, and is a member of the Institute of Electrical and Electronic Engineers (IEEE) and the IESNA Roadway Lighting Committee (RLC). He currently serves as chairman of the RLC Standard Practice Sub-Committee and is a member of the AASHTO Joint Task Force on Roadway Lighting. Mr. Watson is co-author of a chapter on Roadway Lighting in McGraw-Hill's *Highway Engineering Handbook* and of Section 3 of the FHWA Publication FHWA-HI-97-026, *Design Construction and Maintenance of Highway Safety Features and Appurtenances* (NHI Course No.38034). Mr. Watson holds both a Bachelor and a Master of Science degree in Electrical Engineering from Auburn University.

APPENDIX B
AMPLIFYING QUESTIONS

Questions that are numbered were sent to the Europeans.
Questions that are bulleted were of secondary importance.

1. Future Developments

   1.1 What cutting-edge roadway and tunnel lighting research has recently been or is about to be done?

   1.2 We would appreciate a discussion on what you see as the next major roadway and tunnel lighting advancements.

2. Practical Matters of Roadway Illumination Systems

(We would greatly appreciate the benefit of your experience on matters of design, installation, maintenance, and repair of lighting systems.)

   2.1 Design

   2.1.1 What standards or lighting reference documents do you use to determine design requirements, including bibliographic listings? Would it be possible for us to receive a copy of these standards?

   2.1.2 We would appreciate a discussion about the lighting design for roundabouts and other specific geometric features.

   2.1.3 Were optimization studies done to determine your practices regarding mounting height and number/wattage of fixtures for conventional (< 20-m mounting height) and high-mast (>20-m mounting height) lighting? Are copies of such studies available?

   2.1.4 Do you use temporary work zone lighting at roadway construction areas?

   2.1.4.1 In these areas, what standards or requirements do you use to determine the appropriate roadway lighting for the motorist?

   2.1.4.2 What are the requirements for the lighting in the work area for the construction workers? May we have copies of these standards?

   2.1.5 We would appreciate a discussion on any light pollution concerns you may have and actions taken to abate light pollution.

   2.1.6 Please discuss mitigation of specific headlamp characteristics (i.e., sharp cut-off) in lighting designs, especially as applied to a visibility-based design and to the visibility of nonilluminated, retroreflective signs.

- Is an estimate made of the amount of design work required for a particular project? How is this done?

- What resources are committed to lighting?

- Do the people doing the lighting design work have other duties, or do they only work on lighting? What are the qualifications for a lighting designer?

- Does safe stopping distance play a role in the design of speed limitless autobahn lighting designs?

- How do you deal with the differences in elevations of crossroads and ramps?

- What types of low elevation lighting, i.e., curb or rail level, are being used for lighting on structures?

- In the United States, as a safety requirement, we require the poles adjacent to the roadway to "break away" when impacted by an automobile. Do you have similar mandated requirements, or policies, concerning the use of breakaway poles? Are there any mandated requirements, or policies, concerning foundations for breakaway poles?

- How does the wiring system break away? When testing breakaway devices for poles, is the wiring system in place to determine its effect on the breakaway process?

- How is the decision to install lighting on a particular roadway made?

- Do you use warrants for lighting?

- Discuss your warranting conditions for continuous freeway lighting. For complete interchange lighting? For partial interchange lighting?

- Do you use lighting programs that are compatible with CADD? Do you use Intergraph? What other CADD programs are used? Do they handle designs for both luminance and illuminance? CADD application software appears to be limited, especially for luminance design. Are your programs generated in-house or provided by vendors? Who are the vendors?

- Discuss your wiring methods for high-mast and conventional lighting, i.e., circuit parameters (amps, volts, and volt-drop), types of insulation used, locations of fuses.

    2.2   Verification

        2.2.1   Do you have construction acceptance testing and inspection programs?

        2.2.2   How are the design criteria of an installation verified before the customer (state, city, township, etc.) accepts it and pays the contractor/consultant?

            2.2.2.1   What measurement techniques are used to verify the design?

            2.2.2.2   Do you verify calculation methods and determine if field results match calculations (check the amount of error)?

            2.2.2.3   Do you investigate field modifications and how such modifications affect the design?

2.2.2.4   What inspection effort is required? We would appreciate a discussion on your preferences.

- What do you see as the major variables affecting field measurement:

  - For the Illuminance Design Method?

  - For the Luminance Design Method?

- How do you account for different pavement types (concrete, asphalt, aggregate, artificial brighteners), weather conditions (i.e., dampness or dryness of pavement), and wear of pavement, in the luminance readings?

- Bidding

- If you use competitive bidding, how do you account for the differences in photometrics between manufacturers?

- Do you place restrictions on the luminaires as far as efficiency and utilization?

- What are your purchasing processes and procedures?

- Do you buy only the low-bid items?

- Do you set up any long-term contracts with a single vendor for standardization purposes? If so, what is the duration of contract/s?

Operational Issues

- Do you use noncycling HPS lamps? Why?

- What techniques are used to identify and prevent or repair rust and corrosion of lighting pole bases and connections?

- Do you contract out your maintenance? Could we have details? Does it include a monitoring activity?

- Do you have a computerized inventory control?

- How do you keep track of the infrastructure that is out on the street? Manually? Computer?

- What are your emergency response times for weather-related accidents and also for accidents caused by humans?

- Discuss your lighting maintenance policy/procedures for high-speed freeways.

  2.3   Litigation

  2.3.1   How do you protect against litigation if the lighting does not meet standards?

3. Visibility Design Techniques

(The ANSI/IESNA publication *Recommended Practice for Roadway Lighting* has just been revised to include the use of a visibility metric and STV, small target visibility.

Naturally, we are very interested in any experiences you have had using visibility design techniques in your country.)

3.1 Have you conducted studies comparing the results of a visibility-based design to illuminance or luminance-based designs?

3.2 What is the current status of installing roadway lighting systems using a nonuniform luminance pattern based on STV (small target visibility), or close to STV, design principles in your country?

3.3 What percentage of designs used visibility as a basis?

3.4 What standards or lighting reference documents do you use to determine your visibility design requirements, including bibliographic listings?

3.5 Are accident statistics available and do the statistics show an improvement in accident rate or an increase in driver comfort?

3.6 Are any cost figures available comparing these visibility-based designs to more conventional (luminance-based) designs?

3.7 Do you have feedback from drivers, bicyclists, pedestrians, on the visual acceptability (or nonacceptability) of such systems?

3.8 Because it is necessary for computers to be used in calculating visibility levels (VL), discuss the program parameters, such as the equation, size and shape, and reflectance of the target, the contrast of the target, location of calculation points, the effect of vehicle headlights, etc.

3.9 Do you find differences in the resulting lighting systems when different types of targets are used? For example: Would a three-dimensional shape (such as a soccer ball shaped target) yield a different lighting system than one designed around a two-dimensional flat target of similar size and reflectance qualities?

3.10 We would appreciate a discussion on the visibility measurement techniques used to verify design.

3.11 We would appreciate a discussion on the adequacy of lower luminance levels to be used with STV design methods.

   3.11.1 Do you consider headlight contribution in a visibility-based design and how is this affected by different headlamp designs?

3.12 We would appreciate a discussion on the adequacy of STV designed lighting systems in varying (particularly wet or snow) weather conditions.

3.13 We would appreciate a discussion on the various authorities' specifications, inspection, and testing requirements for luminaires used with visibility-based design methods.

   3.13.1 How repeatable do you find the luminaires from a photometric viewpoint? For example, do luminaires installed in different years give the same measurements on the street?

3.14 Is there a difference in driver eye heights used in European STV designs?

4. Luminance Design Techniques

    4.1   How do you account for glare?

- Do you find that luminance designs fit into interchange areas that have varying widths and curved roadways and slower speed crossroads?

- Do you apply luminance to bridge lighting or partial interchange lighting where you may only have a few lights?

- We notice that a design based on luminance tends to push the luminaire out over the roadway, while maintenance desires the luminaire over the roadway's shoulder. Is this an issue and how has this issue been addressed?

- What type of luminaires do you find most useful and why?

- What standard do you use to determine the proper light level and what Candela/m^2 values do you use? What values for Glare ratio?

5. Illumination of Tunnels, High-Mast, Signs, Rest Areas, etc.

    5.1   Tunnels

        5.1.1   We would appreciate a discussion on the various methods of lighting tunnels used in your jurisdiction (pro-beam, counter-beam, symmetrical, light guides, etc.)

        5.1.2   What is your experience with the use of sunscreens before the tunnel portal?

        5.1.3   What standards or lighting reference documents do you use to determine your design requirements?

            5.1.3.1   What design methods are used to determine tunnel lighting levels? (e.g., fixed values, L20, Lseq., snow, atmospheric luminance, exterior contribution in the threshold area)?

        5.1.4   Do you account for glare in tunnel lighting designs?

        5.1.5   How do you determine what is a tunnel and what is an underpass?

            5.1.5.1   How do you light underpasses?

            5.1.5.2   When are underpasses or tunnels not lighted?

        5.1.6   Please discuss your lighting design techniques for very long tunnels, including fixture mechanical characteristics, wiring methods, controls, hypnotic effects, and backup power requirements.

        5.1.7   What has been your experience in the use of induction fluorescent for tunnels, walkways, and bikeways? In particular, what has been the maintenance personnel's experience?

- Are the road design speeds maintained throughout the tunnel or are they reduced?

- What design programs do you have to address tunnel lighting?

- How are your tunnels maintained (i.e., washing, relamping, etc.)?

- Are fire/smoke issues considered in the tunnel equipment? Are materials such as PVC allowed?

- What materials are used in tunnel construction? Are walls and ceilings tile, concrete, shotcrete, other?

- Do you consider fluorescent lighting?

- Are most tunnels luminaires overhead or wall mounted?

- Are tunnel exits provided with increased lighting levels?

   5.2   High-Mast (Mounting Heights > 20 m) Lighting

   5.2.1   What is the experience of using high-mast lighting of roadways and interchanges compared with usual pole heights of 10- to15-m range?

   5.2.2   We would appreciate a discussion on your experiences with mounting heights, number and wattage of fixtures, photometric patterns, and types of lamps for high-mast lighting.

   5.2.2.1   We would appreciate a discussion about the design techniques you use for high-mast lighting (e.g., visibility-based design? uniformity level on high ramps in interchanges? how effectively are grade changes considered?).

   5.3   Signs

   5.3.1   We would appreciate a discussion on the various sign lighting methods used in your jurisdiction (e.g., what light levels (illuminance or illuminance) and uniformities, types of luminaires, light sources, and locations are used?)

   5.3.2   Due to cost and energy concerns, many signs now depend on retroreflective materials and illumination from vehicle headlamps. Given the sharp cutoff of the typical European low-beam pattern, do you experience any problems with proper and timely sign detection, recognition, and legibility?

   5.3.2.1   What are your plans relative to making sure all signs will be detected and read and understood in sufficient time for drivers to take proper action?

- Do you accept lower levels for sign lighting in areas where solar power is used?

5.4 Decorative Lighting

    5.4.1 What types of newer residential and urban street lighting treatments are being used?

    5.4.2 What are your decorative illumination design criteria, (e.g., glare, illuminance, luminance, semi-cylindrical?)

6. Pavement Reflection Factors

    6.1 We would appreciate a discussion about the various reflectance factors used in luminance and visibility-based designs.

    6.2 What is your experience with the effects of R-factors changing due to pavement age, aggregates, rutting, seal coat types, super-elevation, and texturing?

    6.3 Is the reflectance of future roadway treatments considered or controlled (e.g., are any special pavement toppings or cover coats or other techniques being used to control or improve pavement reflectivity?)

        6.3.1 What effect do these have on sky glow?

    6.4 We would appreciate a discussion about variations in off-roadway reflectances under varying weather conditions and how that would impact luminance level requirements.

- Do you consider worst-case condition for dry pavement?
- Do the R-tables adequately describe the roadway's reflectance characteristics?

7. Experience with Counter-Beam vs. Pro-Beam Technology

    7.1 We would appreciate a discussion on accident history for pro-beam, counter-beam, and asymmetric systems.

    7.2 Do you consider using pro-beam or counter-beam on divided highways or tunnels and are there object identification advantages of one vs. the other?

    7.3 We would appreciate a discussion on transition adaptation from pro-beam to counter-beam systems and from HPS/LPS to metal halide systems.

    7.4 We would like to discuss with you the relative position and speed detection of drivers in negative contrast situations (counter-beam).

APPENDIX C
KEY CONTACTS IN HOST COUNTRIES

BELGIUM

R-Tech

Francis Shcreder
Director
Tel: 32 4 224 71 40
Email: gh@rtech.be

Marc Gillet
Director of Lighting Applications
Department
Tel: 32 4 224 71 48
Email: marc.gillet@rtech.be

Ing. Marcel Justin
Lighting Applications Manager
Tel: 32 4 224 71 48
Email: marcel.justin@rtech.be

Jean-Marie Dijon
Tel: 32 4 233 77 47
Email: jm.dijon@swing.be

Ing. Gérard Herman
Project Manager
Tel: 32 4 224 71 40
Email: gh@rtech.be

Schréder

Ir J.-P. Vanhecke
Tel: 32 3 890 66 66
Email: jp.vanhecke@schreder-us.be

Ministry of the Flemish Community

Ir. Jozef Van Ginderachter
Tel: 02 553 72 92
Email: jozefcp.vanginderachter@lin.vlaanderen.be

Ir. Philippe Boogaerts
Tel: 02 553 72 44
Email: philippelt.boogaerts@lin.vlaanderen.be

FINLAND

Finnish National Road Administration (Finnra)

Jukka Isotalo
Director
Tel: 358 0 204 44 2006
Email: jukka.isotalo@tieh.fi

Kari Lehtonen
Research and Design Engineer
Traffic and Road Engineering
Tel: 358 0 204 44 2317
Email: kari.lehtonen@tieh.fi

Arto Tevajärvi
Project Manager
International Affairs
Tel: 358 0 204 44 2032
Email: arto.tevajarvi@tieh.fi

Tuuli Ryhanen
Traffic Management Centre
Tel: 358 0 204 44 8702

SITO Group

Pentti Hautala
Tel: 358 9 476 111
Email: pentti.hautala@sito.fi

Lansitek Contracting

Jukka Salmela
Managing Director
Tel: 358 10 45 57400
Email: jukka.salmela@lansivoima.fi

Helsinki Energy (SEU)

Teemu Rinne

FRANCE

CERTU (Center for Studies on Urban Planning, Transport, Utilities, and Public Construction)

Marc Ellenberg
Deputy Director for Scientific and Technical Coordination
Tel: 33 04 72 74 58 03
Email: ellenber@certu.fr

Robert Charvin
Tel: 33 04 72 74 58 60
Email: rcharvin@certu.fr

Isabelle Vallance
Tel: 33 04 72 74 59 38
Email: Vallance@certu.fr

CETU (Ministry of Equipment, Transport, and Planning)

Jean-Claude Martin
Tel: 04 72 14 34 16
Email: jean-claude.martin@cetu.equipement.gouv.fr

Didier Lacroix
Research Manager
Tel: 33 4 72 14 33 85
Email: didier.lacroix@cetu.equipement.gouv.fr

Philips

Philippe Gandon-Léger
International Product Manager, Outdoor Lighting
Tel: 33 04 78 55 81 99
Email: philippe.gandon-leger@philips.com

Ir. Wout van Bommel
Lighting Design and Application Centre Manager LiDAC Central
Tel: 31 40 27 56314
Fax: 31 40 27 56406
Email: Wout.vanBommel@ehv.lighting.philips.com

Laurent de Ridder
Outdoor Lighting Application Centre Manager
Tel: 33 0 4 72 25 19 91

Ing. G.H.M. Giesbers
Lighting, Design and Application Centre
Tel: 33 0 4 78 55 82 71
Email: gerard.giesbers@philips.com

AFE

Bernard Duval
Tel: 01 45 05 72 80
Email: bduval@feder-eclairage.fr

French National Committee of CIE

Jacques Lecocq
Tel: 33 32 21 48 00
Email: jacques.lecocq@tlgfr.com

Jean Bastie
Email: bastie@cnam.fr

Public Lighting Commission

Christian Remande
Fax: 33 1 48 16 17 89

Paris Laboratory

Françoise Jousse
Email: Francois.Jousse@mairie-paris.fr

Michelle Isaac-Camara
Fax: 33 1 45 80 81 72

Laboratoire Central des Ponts et Chaussées (LCPC)

Corine Brusque
Tel: 33 1 40 43 50 00
Email: Corine.Brusque@lcpc.fr

Vincent Ledoux
Email: Vincent.Ledoux@lcpc.fr

APPENDIX C

Sophie Mosser
Email: Sophie.Mosser@lcpc.fr

Giselle Paulmier
Email: Giselle.Paulmier@lcpc.fr

Eric Dumont
Email: Eric.Dumont@lcpc.fr

Roland Brémond
Email: Roland.Bremond@lcpc.fr

Bernard Jacob
Email: Bernard.Jacob@lcpc.fr

Pierre-Yves Texier
Email: Pierre-Yves.Texier@lcpc.fr

Bernard Mamontoff
Email: Bernard.Mamontoff@lcpc.fr

Laboratoire Régional des Ponts et Chaussées de Lermont-Ferrand

Michèle Colomb
Email: Michele.Colomb@cetelyon.equipement.gouv.fr

Laboratoire Régional des Ponts et Chaussées de Rouen

Alexis Bacelar
Email: Alexis.Bacelar@equipement.gouv.fr

Jacques Cariou
Email: Jacques.Cariou@equipement.gouv.fr

SWITZERLAND

Consultants

Peter Blaser
Tel: 41 31 352 2637
Email: kasi.blaser@bluewin.ch

Hans Meier
Tel: 41 1 736 5429
Email: hans.meier@bd.zh.ch

Lichttechnische Beratung (LIBE)

Werner Riemenschneider
Tel: 056 28 14 35

Elektrizitätswerke des Kantons Zürich (EKZ)

Manfred Jäger
Tel: 01 207 52 60
Email: mjaeger@ekz.ch

Beleuchtungs-Technik AG (BETAG)

Philipp Riemenschneider
Tel: 01 730 77 11

Ingenieur-Unternehmung AG Bern (IUB)

Hans-Rudolf Scheidegger
Tel: 41 0 31 357 11 11
Email: iub.bern@bluewin.ch

Electrowatt-Ekono

Emil Keller
Tel: 411 355 55 55
Email: emil.keller@ewe.ch

Kantonspolizei Zürich

Martin E. Weissert
Tel: 01 247 37 33

Bfu, bpa, upi

Paul Reichardt
Tel: 031 390 22 22
Email: p.reichardt@bfu.ch

Elektrizitätswerk der Stadt Zürich (EWZ)

Jürg Streich
Tel: 01 319 49 01
Email: jurg.streich@ewz.stzh.ch

Peter Schriber
Tel: 01 319 49 02
Email: peter.schriber@ewz.stzh.ch

Martin Bruppacher
Tel: 01 319 49 03
Email: martin.bruppacher@ewz.stzh.ch

THE NETHERLANDS

Ministry of Transport, Public Works, and Water Management

Ir J.W. Huijben
Tel: 030 285 79 82
Email: j.w.huijben@bwd.rws.minvenw.nl

Ton van den Brink
Tel: 31 10 282 59 15
Email: t.d.j.vdbrink@avv.rws.minvenw.nl

Arjen Blacquiere
Tel: 030 285 73 65
Email: a.r.blacquiere@bwd.rws.minvenw.nl

Jitka Usselstijn
Tel: 31 10 282 57 18
Email: j.usselstijn@avv.rws.minvenw.nl

TNO Human Factors Research Institute

Ing. W. Hoekstra
Simulator Specialist
Tel: 03463 5 64 49
Email: wytze@izf.tno.nl

Jeroen H. Hogema
Traffic Behaviour
Tel: 31 346 35 64 40
Email: Hogema@tm.tno.nl

Ing. Johan W.A.M. Alferdinck
Displays, Department of Perception
Tel: 31 346 35 63 11
Email: alferdinck@tm.tno.nl

Noord-Holland

Paul J. Rutte
Tel: 023 514 51 51

APPENDIX D
KEY PAPERS

Throughout the tour, the scanning team received numerous documents from all of the hosting agencies. The documents referred to in this report are listed here.

1. *Color contrast in tunnels* (Kleurcontrast in tunnels) (TNO-report TM-00-C009, in Dutch) by J. W. A. M. Alferdinck, Department of Perception (Displays), TNO Human Factors, Soesterberg, the Netherlands.

2. *Visibility in road lighting; correlation of subjective assessments with calculated values* by J. Lecocq, Thorn Europhane, Les Andelys, France.

3. *Quality criteria for road lighting: luminance and uniformity levels? Or visibility?* by Jean-Marie Dijon and Laurent Maldague of R-Tech S.A. Belgium.

4. *Tunnel of Wevelgem; comparison and tests of symmetrical, counter-beam and pro-beam systems* by Jean-Marie Dijon and P. Winkin of R-Tech S.A. Belgium.

5. *Laboratory experimental study of the influence of environmental complexity on the detection of various targets* by G. Paulmier, C. Brusque, V. Carta, and V. Nguyen, LCPC; Paris, France.

6. *Evaluation of the photometric characteristics of porous asphalts* by Corinne Brusque, Jean Peybernard, LCPC; Paris, France.

7. *Study of road surface photometric properties by numerical simulations* by T. Rondeau, C. Brusque, LCPC, Paris, France; and N. Noe and B. Peroche of Ecole des Mines de St-Etienne Centre Simade, St-Etienne, France.

8. *Black window method of short tunnel lighting design* by J. W. Huijben and F. de Roo, Engineering Department, Rijkwaterstaat, the Netherlands.

9. *Dynamic public lighting*, Ministry of Transport, Public Works, and Water Management; P.O. Box 1031; 3000 BA Rotterdam; the Netherlands.

APPENDIX E
OUTREACH ACTIVITIES IN 2000

CONFERENCES AND SYMPOSIUMS

Name	Time	Presenter(s)
TRB Visibility Symposium	May 15-16	Havard/Arens
Kansas State Roadway Design Course	May	Hasson
AASHTO Annual Meeting on Traffic Eng.	June	Balu
IMSA	July	Balu
ITE Annual Meeting	August 6-9	Burkett
IESNA – RLC Meeting	July 29	Havard
Street & Area Lighting Conference	September	Havard
Local Wisconsin IESNA Chapter	September	Balu
AASHTO Annual Meeting	October	Hasson/Burkett
APWA Annual Meeting		Hasson
TRB Annual Meeting	January 2001	Burkett

ARTICLES AND PAPERS

Public Roads
IESNA LD&A Magazine
AASHTO Journal
ITE Journal
Public Works Magazine

TEST BEDS FOR POSSIBLE DEMONSTRATIONS

Alabama DOT
Pennsylvania DOT
Texas DOT
Wisconsin DOT
Smart Road - Virginia

ENDNOTES

[1] Hockey, E. and McGee, H. *Minimum Traffic Sign Retro-reflectivity Guidelines: The United States Experience,* presented at 3rd African Road Safety Congress, Pretoria, South Africa, April 1997.

[2] ANSI/ IESNA RP-8-00 "American National Standard Practice for Roadway Lighting."

[3] Dijon, Jean-Marie. *Quality Criteria for Road Lighting: Luminance and Uniformity Levels? Or Visibility?"* IR or R-Tech S.A., Belgium.

[4] Dijon, Jean-Marie. *Quality Criteria for Road Lighting: Luminance and Uniformity Levels? Or Visibility?* IR or R-Tech S.A., Belgium.

[5] Hockey, E. and McGee, H. *Minimum Traffic Sign Retro-reflectivity Guidelines: The United States Experience,* presented at 3rd African Road Safety Congress, Pretoria, South Africa, April 1997.

www.ingramcontent.com/pod-product-compliance
Lightning Source LLC
Chambersburg PA
CBHW081842170526
45167CB00007B/2876